REVISION GUIDE

Eastern Caribbean Primary

Mathematics

Dr Jeffrey Blaize

Advisor: Colin Cumberbatch

CONTENTS

HOW THIS BOOK IS ORGANISED

Eastern Caribbean Primary Mathematics Revision Guide is a complete revision package designed to help students gain the knowledge and skills needed for success with end-of-primary Mathematics examinations.

The content has been carefully ordered into 10 chapters, and broken down into sub topics, written to meet the specifications of Mathematics examinations throughout the Eastern Caribbean, including the BSSEE. This fun, easy-to-work-through Revision Guide and CD-ROM package makes revision effective and less stressful with the following features:

'What you need to know' offers succinct summaries of crucial facts for revision

Each chapter is clearly introduced and includes objectives to make revision more focused

Each topic heading is followed by 'Points to Remember', presenting crucial content in manageable portions

Each chapter is packed with examples and their worked solutions

Each topic is broken up with a range of Exercises, with solutions included at the back of the book

'Now Try it Out' boxes end each chapter, offering additional practice through activities and games

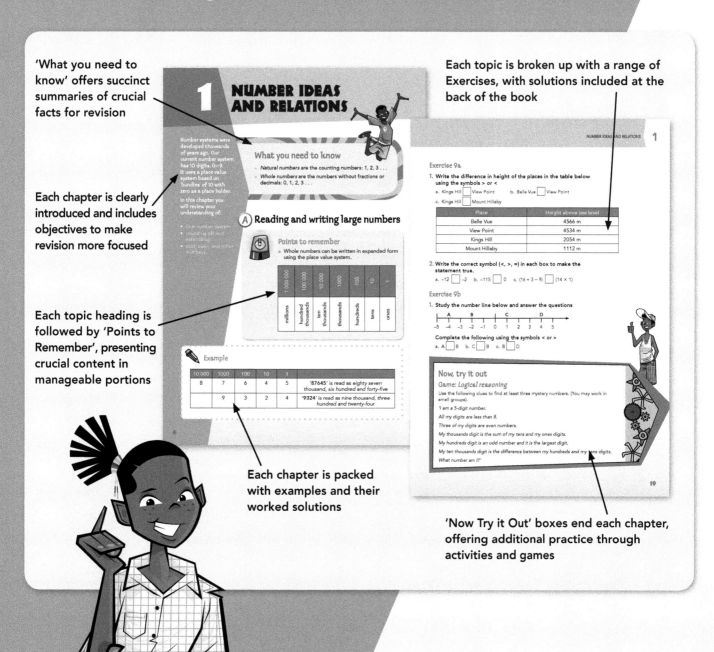

15 multiple-choice and 5 short answer questions are included at the end of each chapter, presented in the style of exam papers. Answers are included at the back of the book.

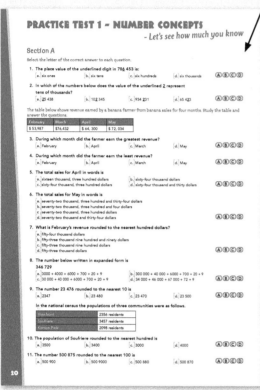

Features of the student CD-ROM

Revision is made even more effective and stress-free with the fun and interactive accompanying CD-ROM. It enhances learning through interactive activities and additional practice opportunities. More exam revision can be found with multiple-choice questions, crosswords and matching exercises help to reinforce key concepts and printable worksheets offer extension activities for both students and teachers.

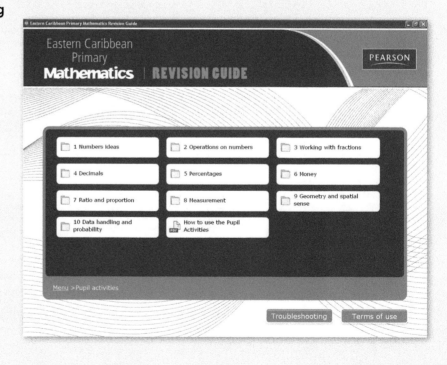

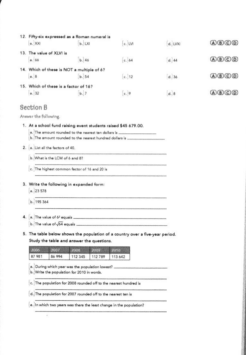

'When I go into the exam I want to feel confident, calm and well prepared. How do I do this?'

Exam guidance and tips

* Do plenty of revision and practice in the weeks before the exam. There are practice papers at the end of each chapter in this book, as well as additional exam-style questions on the CD-ROM.

* Prepare everything you need for the exam (pens and pencils, etc.) the night before. Make sure you have the right kind of pencil for shading the answers on the answer sheet.

* Get a good night's sleep before an exam. Being bright and lively during the exam is more important than what you can learn by studying late the night before.

* Get up early and have a good breakfast to be at your best for the exam.

* Go to school in good time. If you have to rush to get to the exam you might panic.

* Keep calm before and during the exam. If you start to panic, close your eyes and take four or five long, slow, deep breaths to calm yourself down. Tell yourself how much you know. If you are worried by a question you cannot do, remind yourself of all the others that you can do well. Nobody expects you to get 100%.

* Don't rush to start answering questions. Have a quick glance over the whole paper to see how much there is to do.

* Read questions carefully before answering. Make sure you know what you have to do before you put pen to paper.

* For multiple-choice questions:
    - Read the sentence or question carefully.
    - Look carefully at all the choices. In many cases you will see the correct answer straightaway.
    - If you are not sure of an answer, use a process of elimination. You will realise that one or two of the answers are certainly incorrect - cross these out. Now look carefully at the remaining answers and make a decision.
    - If you find you cannot answer a question, leave it and come back to it at the end. Do not waste time getting upset because you find something too difficult. You may find the answer becomes clear when you look at it again later.

* For short-answer questions:
    - Look quickly over the questions to see roughly what kind of information they are asking about.
    - Read the text carefully before attempting to answer the questions.

* Work steadily and carefully through the paper. Try to reach the end a few minutes before the time is up. You can then go back to any questions you have left or check your answers. (Or if you have finished everything and are confident with your answers - take a minute to feel good about yourself!)

Good luck!

Other titles in this series:

Eastern Caribbean Primary English Revision Guide & CD-ROM: 9781408263266
Please visit www.pearsoncaribbean.com/primaryrevision to find out more.

1 NUMBER IDEAS AND RELATIONS

Number systems were developed thousands of years ago. Our current number system has 10 digits, 0—9. It uses a place value system based on 'bundles' of 10 with zero as a place holder.

In this chapter you will review your understanding of:

- our number system
- rounding off and estimating
- odd, even, and other numbers.

What you need to know

- *Natural numbers* are the counting numbers: 1, 2, 3 . . .
- *Whole numbers* are the numbers without fractions or decimals: 0, 1, 2, 3 . . .

(A) Reading and writing large numbers

Points to remember

- Whole numbers can be written in expanded form using the place value system.

1 000 000	100 000	10 000	1000	100	10	1
millions	hundred thousands	ten thousands	thousands	hundreds	tens	ones

 Example

10 000	1000	100	10	1	
8	7	6	4	5	'87645' is read as *eighty seven thousand, six hundred and forty-five*
	9	3	2	4	'9324' is read as *nine thousand, three hundred and twenty-four*

Exercise 1a

1. **What is the place value of the underlined digit in:**

   a. 6_7_8   b. 2_8_4 931?

2. **The chart below shows cash sales of Courts West Indies Ltd for the first six months of 2010.**

January	February	March	April	May	June
$508 793	$180 920	$767 395	$618 092	$1,000 000	$217 005

   a. Write in words the total sales for the month of February.

   b. During which month did the company receive six hundred and eighteen thousand and ninety-two dollars in sales?

3. **The car licence plate shown has five digits.**

   a. What is the value of the digit in the thousands place?

   b. In what position is the digit 5?

   PA 03584

4. **Write down the missing number to complete each equation.**

   a. 365 211   =   300 000 + _____ + 5000 + 200 + 10 + 1

   b. 107 484   =   _____ + 7000 + 400 + 80 + 4

   c. 18 647   =   10 000 + 8000 + 600 + _____ + 7

Exercise 1b

1. **Write each of these numbers on the number line shown here.**

   a. 18 000   b. 24 000

   c. 15 000   d. 28 000

2. **How many 100s make 10 000?**

3. **How many 10 000s make 100 000?**

4. **A man has a lottery ticket with the following numbers.**

   His ticket is almost identical to the winning ticket except for the fact that the winning ticket has the digit 3 in the ten thousands column and the digit 1 in the tens column. What is the full combination of digits in the winning ticket?

   LOTTERY $

   $$ SATURDAY $

   1 7 6 4 5 2 9

Now, Try it out

How long would it take you to spend a million dollars, if you spent,

a. $1000 each day   b. $500 each day

c. $200 each day   d. $100 each day?

(B) Using powers of ten

Points to remember

■ You can use powers of ten to show the expanded form of a number.

$10\ 000 = 10^4$

$1000 = 10^3$

$100 = 10^2$

$10 = 10^1$

$1 = 10^0$

eg 1 —   In the expression 10^4, the number 10 is called the *base*.

The number 4 is called the *exponent*.

10^4 is read as '10 to the 4th power' or '10 to the power of 4'.

The exponent tells how many tens to multiply.

So $10^4 = 10 \times 10 \times 10 \times 10 = 10\ 000$

eg 2 — 15×10^3 is read as '15 × 10 to the 3rd power' or '15 × 10 to the power of 3'

so $15 \times 10^3 = 15 \times 10 \times 10 \times 10 = 15\ 000$

Example

Write the number 75 403 in expanded form.

Solution

10 000	1000	100	10	Ones
7	5	4	0	3
$(7 \times 10\ 000)\ +$	$(5 \times 1000)\ +$	$(4 \times 100)\ +$	$(0 \times 10)\quad +$	(3×1)
$(7 \times 10^4)\qquad +$	$(5 \times 10^3)\quad +$	$(4 \times 10^2)\ +$	$(0 \times 10^1)\quad +$	(3×1)

Exercise 2a

1. **Show the expanded form for:**

   a. 99 234   b. 745 391   c. 1 863 402   d. 1 243 421

2. **Write each number using an exponent (power of 10).**

   a. 1000 =   b. 10 000 =   c. 12 000 =   d. 14 000 =

Exercise 2b

1. **Complete the table.**

Expanded form	Power of 10	Standard notation
$(7 \times 10000) + (5 \times 1000) + (4 \times 100)$ $+ (3 \times 1)$	$(7 \times 10^4) + (5 \times 10^3)$ $+ (4 \times 10^2) + 3$	75 403
1 000 000 + 800 000 + 600 + 70 + 4		
20 000 + 3000 + 500 + 70 + 8		
$(1 \times 10\ 000) + (4 \times 1000) + (5 \times 100)$ $+ (3 \times 10) + (3 \times 1)$		

2. **Write these in standard notation.**

   a. 25×10^3 b. 20×10^2

   c. 15×10^4 d. 12×10^3

Now, try it out

Select 3 six-digit numbers. Rewrite each of the numbers in expanded form. Now write each of the numbers using powers of ten.

© Rounding off

Points to remember

- *Rounding off* means re-writing a number with a given degree of accuracy to the nearest indicated value.
- We usually round the number 5 and above upwards and numbers less than 5 downwards.

 ### Example

Round 792 to the nearest 10 and to the nearest 100.

Solution

The tens digit is 9 and the ones digit after it is 2. So 792 is nearer to 790 than 800 and we write '792 to the nearest 10' is 790.

The hundreds digit is 7 and the tens digit after it is 9 so 792 is nearer to 800 than 700 and we write '792 to the nearest hundred' is 800.

Exercise 3a

1. **Round 31 486 to the nearest:**

   a. ten   b. hundred   c. thousand

2. **The annual salary for the Commissioner of Police is $78 600.**

   What would this figure be to the nearest thousand dollars?

3. **Round 457 548 to the nearest hundred.**

Exercise 3b

1. **Use the table to answer the following questions.**

   a. Which island has a population closest to 1 000 000?

   b. Which island has a population closest to 70 000?

   c. Which island has a population closest to 90 000?

   d. Which two islands would have the same population, if rounded off to the nearest 1000?

   e. What is the population of Barbados to the nearest ten?

Island	Population
St Lucia	917 257
St Kitts and Nevis	89 569
Antigua & Barbuda	110 256
Grenada	90 245
Dominica	70 245
Barbados	255 203

St Lucia

St Kitts/Nevis

Antigua

Grenada

Dominica

Barbados

2. Round off the following numbers to the nearest 100.

    a. 456 732   b. 178 234   c. 34 412

Now, Try it out

There were 19 879 spectators watching a cricket match. Round off the number of spectators to the nearest: 10 000   1000   100.

(D) Roman numerals

Points to remember

- In the *Roman numeral* system, letters are used to represent numerals:
  I = 1   V = 5   X = 10   L = 50   C = 100   D = 500   M = 1000
- If a letter with a smaller numeral value appears to the right of another letter, the numeral is added.
- If a letter with a smaller numeral value appears to the left of another letter, the numeral is subtracted.

 Example

What are the equivalent Roman numerals for the following?

4, 6, 25, 40

Solution

Number	Equivalent	Roman numeral
4	5 – 1	IV
6	5 + 1	VI
25	10 + 10 + 5	XXV
40	50 – 10	XL

Exercise 4a

1. What number is represented by each of the following Roman numerals?

    a. LI   b. LIX   c. XIX   d. LIV   e. XLIV

2. Write Roman numerals for the following numbers.

    a. 29   b. 35   c. 45   d. 49

Exercise 4b

1. How old is each of the students below?

   a. XIII _____yrs   b. IX _____yrs   c. XIV _____yrs

2. The following table shows the ages of various teachers at school. Complete the table.

Teacher	Age	Age in Roman numerals
Mr Austrie	44 years	
Miss Francis		XXIV
Mr Jno Baptiste	43 years	
Ms Dailey		LIV
Mr Joseph	23 years	
Miss Laville	31 years	

Now, Try it out

Work in teams of five. Draw a table inserting the names of each member of the group. Find the age of each member and insert the age in a separate column. Now write each age in Roman numerals using a third column.

E Odd, even, prime and composite numbers

Points to remember

- *Odd numbers* are numbers that cannot be divided by 2 exactly. They end with 1, 3, 5, 7 or 9.
- *Even numbers* are numbers divisible by 2 and end with 0, 2, 4, 6 or 8.
- *Prime numbers* are numbers greater than 1 having just two factors, 1 and the number itself.
- *Composite numbers* are numbers greater than 1 that have more than two different factors.

Exercise 5a

1. List all the factors of each number and then complete the chart using ticks (✔).

Number	Factors	Odd number	Even number	Prime number
23				
48				
67				
70				

2. What is the smallest prime number?

3. Which numbers between 10 and 40 are prime numbers?

Exercise 5b

1. **Study the numbers from 1 to 50 listed.**
   a. List all the even numbers.
   b. List all the odd numbers.
   c. List all the multiples of 9.

2. **Write six prime numbers between 50 and 100.**

3. **Identify the multiples of 7 between 50 and 100.**

4. **Identify the multiples of 12 between 84 and 144.**

```
 1  2  3  4  5  6
 7  8  9  10 11
12 13 14 15 16
17 18 19 20 21
22 23 24 25 26
27 28 29 30 31
32 33 34 35 36
37 38 39 40 41
42 43 44 45 46
   47 48 49 50
```

Now, try it out

You will need small number cards from 1–50 . You are required to work in pairs. Place all numbers on your desks on the reverse side. Each student takes turns at revealing one number while the other student indicates whether the number is prime or composite. One point is awarded for each correct answer.

(F) Factors, HCF

Points to remember

- Every whole number has at least two factors, 1 and itself.
- A *factor* of a number is a number that divides into it without leaving a remainder. For example, the factors of 36 are 1, 2, 3, 4, 6, 9, 12, 18 and 36.
- The *highest common factor* (HCF) of a set of numbers is the largest number dividing each number in the set without leaving a remainder.
- Prime factors are the factors of a given number which are also prime. For example, the factors of 12 are: 1, 2, 3, 4, 6 and 12. Of these 2 and 3 are also prime. Hence, they are prime factors of 12.

Example 1

Express 36 as a product of its prime factors.

Solution

Prime factorization

$$36 \div 2$$
$$18 \div 2$$
$$9 \div 3$$
$$3 \div 3$$
$$1$$

$36 = 2 \times 2 \times 3 \times 3$

or:

Using a tree diagram

$36 = 2 \times 2 \times 3 \times 3$

or:

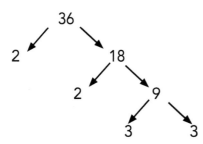

Example 2

Find the highest common factor (HCF) of 12 and 36.

Solution

Factors of 12: 1, 2, 3, 4, 6, 12

Factors of 36:
1, 2, 3, 4, 6, 9, 12, 18, 36

HCF: 12

or:

$12 = 2 \times 2 \times 3$
$36 = 2 \times 2 \times 3 \times 3$

HCF of 12 and 36 is
$2 \times 2 \times 3 = 12$

Exercise 6a

1. **Find the HCF of:**

    a. 27 and 36   b. 45 and 60   c. 15, 18 and 24

2. **Identify all the factors which are common to the following three numbers.**

    30, 40, 54

3. **Identify the remaining factors to complete the blank spaces.**

    a. Number ___24___ :   Factors   1, 2, 3, 4, 6, 8, _____ , _____ .

    b. Number ___18___ :   Factors   1, 2, 3, 6, _____ , _____ .

    c. Number ___16___ :   Factors   1, 2, 4, _____ , _____ .

Exercise 6b

1. **Write these numbers as products of their prime factors.**

    a. 24   b. 36   c. 45   d. 60

2. **Complete the following table.**

Number	Factors
30	
44	
50	
75	

Now, try it out

Find the largest number which when divided into 34 and 48 will leave a remainder of 6 in each case.

G Multiples, LCM

Points to remember

- A multiple of a number is a number that is exactly divisible by that number.
- The *lowest common multiple* (LCM) of a set of numbers is the smallest number that is divisible by every number in the set.

Example

Find the LCM of 24 and 72.

Solution

Multiples of 24: 24, 48, (72), 96, 120, (144), …
Multiples of 72: (72), (144), …
Lowest common multiple (LCM) is 72

or: or:

2	24, 72
2	12, 36
3	6, 18
2	2, 6
3	1, 3
	1, 1

Prime factorization

24 ÷ 2   72 ÷ 2
12 ÷ 2   36 ÷ 2
 6 ÷ 2   18 ÷ 2
 3 ÷ 3   9 ÷ 3
   1   3 ÷ 3
          1

LCM of 24 and 72 = 2 × 2 × 2 × 3 × 3 = 72

= 2 × 2 × 3 × 2 × 3 = 72

Exercise 7a

1. **Find the LCM of:**

   a. 12 and 15   b. 30 and 75   c. 6, 14 and 21.

2. **Write five multiples which are common to 3 and 6.**

3. **Identify two multiples which are common to 4, 6 and 8.**

Exercise 7b

1. **Find the LCM of the following numbers.**

   a. 4, 6, 12   b. 5, 10, 15   c. 2, 8, 12   d. 3, 6, 9

2. **List the first five multiples of each of these numbers.**

   a. 4   b. 6   c. 9   d. 12

3. **Complete the table below.**

Number	First five multiples
6	
8	
10	
12	

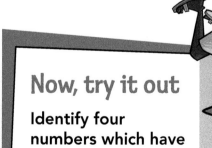

Now, try it out

Identify four numbers which have 48 as a multiple.

H Squares and square roots

Points to remember

- A number is squared when it is multiplied by itself.
- 4^2 is read 'four to the 2nd power' or '4 squared'.
- The little number on the top or the *raised number* is called the *exponent* or *power*. It expresses the number of times you multiply the larger number by itself.
- Finding the square root is the opposite operation to squaring a number.
- The symbol for square root is $\sqrt{}$.
- You will recognise some square roots, e.g. $\sqrt{144} = 12$. For others you will need a scientific calculator.

Example

What is the square root of 81?

Solution

$9^2 = 81$. The square root of 81 is 9 or $\sqrt{81} = 9$

Exercise 8a

1. **What is the value of:**

   a. 5^2 b. 12^2 c. 11^2 d. 10^2?

2. **Find the square root of:**

   a. 64   b. 16   c. 36

3. **Find the value of:**

   a. $\sqrt{25}$ b. $\sqrt{144}$ c. $\sqrt{64}$

Exercise 8b

1. Complete the table below.

Number	Representation	Value
9^2	9×9	81
3^2		
	5×5	
		64
	12×12	
		121

Now, try it out

Find the sum of the following
$8^2 + 9^2 + 12^2 =$

2. Find the value of the following:

a. 7^2 b. 8^2 c. 6^2

① The number line and integers

Points to remember

- *Integers* are positive and negative whole numbers.
- *Positive numbers* are to the right of 0 on the number line.
- *Negative numbers* are to the left of 0 on the number line.
- 0 is neither negative nor positive.
- Positive numbers get larger as they move away from 0; that is, going from left to right.
- Negative numbers get smaller as they move away from 0; that is, going from right to left.
- The symbol > means 'is greater than'
  < means 'is less than'
  ≥ means 'is greater than or equal to'
  ≤ means 'is less than or equal to'.

Example

Write the correct symbol (<, >, =) in the boxes to make the statement true.

Solution

6 ☐ 18

6 < 18

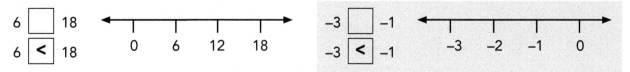

0   6   12   18

−3 ☐ −1

−3 < −1

−3   −2   −1   0

Exercise 9a

1. **Write the difference in height of the places in the table below using the symbols > or <**

   a. Kings Hill ☐ View Point   b. Belle Vue ☐ View Point

   c. Kings Hill ☐ Mount Hillaby

Place	Height above sea level
Belle Vue	4566 m
View Point	4534 m
Kings Hill	2054 m
Mount Hillaby	1112 m

2. **Write the correct symbol (<, >, =) in each box to make the statement true.**

   a. -12 ☐ -2 b. -115 ☐ 0 c. $(16 + 3 - 9)$ ☐ (14×1)

Exercise 9b

1. **Study the number line below and answer the questions**

   Complete the following using the symbols < or >

   a. A ☐ B   b. C ☐ B   c. B ☐ D

Now, try it out

Game: *Logical reasoning*

Use the following clues to find at least three mystery numbers. (You may work in small groups).

'I am a 5-digit number.

All my digits are less than 8.

Three of my digits are even numbers.

My thousands digit is the sum of my tens and my ones digits.

My hundreds digit is an odd number and it is the largest digit.

My ten thousands digit is the difference between my hundreds and my tens digits.

What number am I?'

PRACTICE TEST 1 – NUMBER CONCEPTS

- Let's see how much you know

Section A

Select the letter of the correct answer to each question.

1. The place value of the underlined digit in 786 453 is:

   [a.] six ones   [b.] six tens   [c.] six hundreds   [d.] six thousands   Ⓐ Ⓑ Ⓒ Ⓓ

2. In which of the numbers below does the value of the underlined 2 represent tens of thousands?

   [a.] 25 438   [b.] 102 345   [c.] 934 231   [d.] 65 423   Ⓐ Ⓑ Ⓒ Ⓓ

The table below shows revenue earned by a banana farmer from banana sales for four months. Study the table and answer the questions.

February	March	April	May
$53 987	$76 432	$64 300	$72 034

3. During which month did the farmer earn the greatest revenue?

   [a.] February   [b.] April   [c.] March   [d.] May   Ⓐ Ⓑ Ⓒ Ⓓ

4. During which month did the farmer earn the least revenue?

   [a.] February   [b.] April   [c.] March   [d.] May   Ⓐ Ⓑ Ⓒ Ⓓ

5. The total sales for April in words is

   a. sixteen thousand, three hundred dollars   b. sixty-four thousand dollars

   c. sixty-four thousand, three hundred dollars   d. sixty-four thousand and thirty dollars   Ⓐ Ⓑ Ⓒ Ⓓ

6. The total sales for May in words is

   a. seventy-two thousand, three hundred and thirty-four dollars

   b. seventy-two thousand, three hundred and four dollars

   c. seventy-two thousand, three hundred dollars

   d. seventy-two thousand and thirty-four dollars   Ⓐ Ⓑ Ⓒ Ⓓ

7. What is February's revenue rounded to the nearest hundred dollars?

   a. fifty-four thousand dollars

   b. fifty-three thousand nine hundred and ninety dollars

   c. fifty-three thousand nine hundred dollars

   d. fifty-three thousand dollars   Ⓐ Ⓑ Ⓒ Ⓓ

8. The number below written in expanded form is

   346 729

   a. 3000 + 4000 + 6000 + 700 + 20 + 9   b. 300 000 + 40 000 + 6000 + 700 + 20 + 9

   c. 30 000 + 40 000 + 6000 + 700 + 20 + 9   d. 34 000 + 46 000 + 67 000 + 72 + 9   Ⓐ Ⓑ Ⓒ Ⓓ

9. The number 23 476 rounded to the nearest 10 is

   [a.] 2347   [b.] 23 480   [c.] 23 470   [d.] 23 500   Ⓐ Ⓑ Ⓒ Ⓓ

10. In the national census the populations of three communities were as follows.

Viewfront	2356 residents
Soufriere	3457 residents
Kenton Park	2098 residents

   The population of Soufriere rounded to the nearest hundred is

   [a.] 3500   [b.] 3400   [c.] 3000   [d.] 4000   Ⓐ Ⓑ Ⓒ Ⓓ

11. The number 500 875 rounded to the nearest 100 is

   [a.] 500 900   [b.] 500 9000   [c.] 500 880   [d.] 500 870   Ⓐ Ⓑ Ⓒ Ⓓ

12. Fifty-six expressed as a Roman numeral is

    a. XXI   b. LXI   c. LVI   d. LVXI   Ⓐ Ⓑ Ⓒ Ⓓ

13. The value of XLVI is

    a. 66   b. 46   c. 64   d. 44   Ⓐ Ⓑ Ⓒ Ⓓ

14. Which of these is NOT a multiple of 6?

    a. 8   b. 54   c. 12   d. 36   Ⓐ Ⓑ Ⓒ Ⓓ

15. Which of these is a factor of 16?

    a. 32   b. 7   c. 9   d. 8   Ⓐ Ⓑ Ⓒ Ⓓ

Section B

Answer the following.

1. At a school fund raising event students raised $45 679.00.

    a. The amount rounded to the nearest ten dollars is _____

    b. The amount rounded to the nearest hundred dollars is _____

2. a. List all the factors of 40.

    _____

    b. What is the LCM of 6 and 8?

    _____

    c. The highest common factor of 16 and 20 is

    _____

3. Write the following in expanded form:

    a. 23 578

    _____

    b. 195 364

    _____

4. a. The value of 6^2 equals _____

    b. The value of $\sqrt{64}$ equals _____

5. The table below shows the population of a country over a five-year period. Study the table and answer the questions.

2006	2007	2008	2009	2010
87 981	86 994	112 345	112 789	113 642

    a. During which year was the population lowest? _____

    b. Write the population for 2010 in words.

    _____

    c. The population for 2008 rounded off to the nearest hundred is

    _____

    d. The population for 2007 rounded off to the nearest ten is

    _____

    e. In which two years was there the least change in the population?

    _____

OPERATIONS ON NUMBERS

The four operations are used by almost everyone for tasks ranging from simple day-to-day calculations to advanced scientific calculations. Operations involve studying quantities and manipulating numbers. The four operations are addition, subtraction, multiplication and division.

In this chapter you will review and reinforce your understanding of:

- the four operations (+, −, ×, ÷)
- the order to use operations in a calculation
- the distributive, commutative and associative laws.

What you need to know

The key words associated with the four operations are:

* *Addition*: sum, total, altogether, plus, increase by, addend
* *Subtraction*: difference, minus, less, decrease by, subtractend
* *Multiplication*: multiplier, product, times, multiplicand
* *Division*: divisor, dividend, quotient, per, share, ratio.

Addition	215	+	75	= 290 *sum*
Subtraction	290	−	75	= 215 *difference*
Multiplication	67	×	10	= 670 *product*
Division	670	÷	10	= 67 *quotient*

(A) Addition and subtraction

Points to remember

■ Addition and subtraction are opposite operations; one is the inverse of the other, e.g. 8 + 5 = 13 and 13 − 5 = 8.

■ When adding and subtracting numbers, *place value* is essential. Ensure that you correctly align each number.
A = 10 695 m,   B = 6494 m,
C = 9320 m,   D = 9032 m,
E = 5743 m,   F = 4483 m

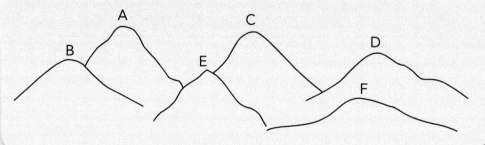

Example 1

What is the difference in height between the lowest and highest peaks?

Solution

Highest peak A   10 695 m

Lowest peak F   4483 m

Difference   subtract 4483 m from 10 695 m

First you can estimate your answer before doing the actual calculation:

10 695 ≈ 10 700

  4483 ≈   4500

Estimate: 10 700 – 4500 = 6200

Then you can do the calculation:

10^4	10^3	10^2	10^1	Ones
1	0	6	9	5
–	4	4	8	3
6	2	1	2	

The difference in height between peak A and peak F is 6212 m.

Exercise 1a

1. **The information in the table shows Mrs Jones' water meter reading over a period of six months. Use this information to answer the following questions.**

1st reading month 1	2nd reading month 2	3rd reading month 3	4th reading month 4	5th reading month 5	6th reading month 6
10 237	10 348	10 451	10 547	10 599	10 672

a. During which month did Mrs Jones use the most water?

b. How many units of water were used over the six-month period?

c. Find the difference between the 2nd and 4th readings.

2. Complete the table.

	Problem	Estimated answer	Calculated answer
a.	243 + 126 + 60		
b.	1562 + 483 + 217		
c.	245 + 314 + 723		
d.	8794 − 2613		
e.	2603 − 768		

3. John's bank account is $40 overdrawn (that is −$40). He pays in his wages of $150. How much has he in his account now?

4. The highest point in a country is 1478 m above sea level. If the difference between this point and a second point is 650 m, find the height of the second point.

5. A soda factory loaded 9001 crates of soda on Monday and 7509 on Tuesday. If the total number of crates shipped over the three-day period was 20 510, how many crates were shipped on Wednesday?

Exercise 1b

1. The following table represents election votes cast in a constituency on polling day.

Polling District	Votes cast	Spoilt ballots
A-101	392	5
B-102	1603	0
C-103	7204	13
D-104	999	2

   a. How many more votes were cast in Polling District C-103 than in B-102?

   b. What is the total number of votes cast in Polling Districts A-101 and D-104 together?

   c. What is the difference between the numbers of votes cast in Polling District B-102 and those cast in D-104?

   d. Find the total number of votes which were cast in the constituency at that election.

2. At a school graduation, the Student of the Year 2010 received a cash gift of $2450.00. The Student of the Year for the previous year received $1850.00. How much more did the Student of the Year 2010 receive?

3. a. Which two numbers in the box below have a sum of 353?
   b. Which two numbers have a difference of 71?

| 3 | 7 | 78 | 275 | 1000 |

4. Samuel had $7723.00 on his savings account. He donated $300.00 towards an earthquake relief fund and $1000.00 towards his favourite charity. What is Samuel's balance?

Now, try it out

Complete these addition magic squares. Remember the sum of the numbers is found outside of the box.

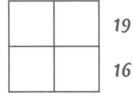

a.

8	___	14
___	4	11
15	10	

b.

		19
		16
15	20	

B Multiplication and division

Points to remember

- *Multiplication and division* are opposite operations: e.g. 4 × 6 = 24 and 24 ÷ 4 = 6.
- Multiplication can be seen as repeated addition: 2 × 3 is the same as 2 + 2 + 2.

You should always give your answer in the simplest form unless otherwise advised.

Example 1

Work out 124 × 36

Solution

```
      124
  ×    36
     3720
      744
     4464
```

 Example 2

Work out 1836 ÷ 24.

Solution

```
        76
  24) 1836
      168
      156
      144
       12
```

The answer is 76 remainder 12 or $76\frac{12}{24} = 76\frac{1}{2} = 76.5$

Exercise 2a

Work out the following:

1. Karen earned $4620 last year. How much did she earn per month?

2. A basket can hold 32 mangoes. How many mangoes can 51 baskets hold?

3. In St Mary's school there are 396 children. How many soccer teams of 11 players can be formed?

4. Work out

   a. 68 × 59 =

   b. 630 ÷ 15 =

5. Four students are playing basketball. Two are of the same height; the other two are 165 cm and 150 cm tall. If the combined height of all four students is 655 cm, what are the heights of the other two students?

Exercise 2b

1. At a cricket match, the opening batsman scored 148 runs before he was bowled out. If the captain scored half as many runs, what was the captain's score?

2. How many guests can be seated at a theatre containing 116 rows of chairs if there are 7 chairs in each row?

3. A farmer was loaned $40 000.00 with no interest towards the purchase of a vehicle. One condition of the loan is that the farmer repays $5000.00 each year. How long would it take the farmer to clear the loan?

4. Solve these problems.

   a. 1620 ÷ 45 =

   b. 540 ÷ 12 =

   c. 1248 ÷ 16 =

Now, try it out

Complete the pattern.

_____ × 5 = 45

_____ × 5 = 450

_____ × 5 = 4500

_____ × 5 = 45 000

_____ × 5 = 450 000

(C) Multiplying and dividing strategies

Points to remember

- To *multiply by powers* of 10 (10, 100, 1000, etc.) you need to shift the decimal point one step right each time you multiply by ten.
- 79 × 100 = 7900 ⟶ add two zeros to the right of the number
  0.023 × 1000 = 23 ⟶ shift the position of the decimal point three places to the right

- To *divide by powers* of 10, you need to shift the decimal point one step to the left each time you divide by 10.
  894 ÷ 100 = 8.94 ⟶ shift the decimal point two places to the left

  0.902 ÷ 10 = 0.0902 ⟶ shift the decimal point one place to the left

Exercise 3a

1. **Calculate:**

   a. 279 × 1000   e. 208 × 99

   b. 8462 ÷ 100   f. 635 × 101

   c. 1968 × 25   g. 278 × 125

   d. 656 × 75   h. 85 × 50

Exercise 3b

1. **Complete the following:**

   a. (   ) ÷ 80 = 4   b. (   ) = 80 × 4

   c. 80 × (   ) = 320

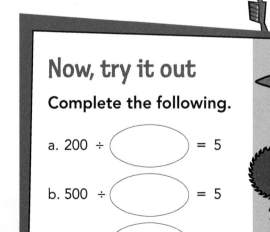

Now, try it out

Complete the following.

a. 200 ÷ (   ) = 5

b. 500 ÷ (   ) = 5

c. 19 × (   ) = 190

(D) Properties of basic mathematical operations

Points to remember

■ *Properties* of addition and subtraction can help you find the *sum*, or *product* of numbers.

Property	Description	Example
Commutative	The order does not make any difference when adding or multiplying two numbers.	$a + b = b + a, a \times b = b \times a$
Associative	The order of grouping does not make any difference when adding or multiplying three numbers.	$(2 + 3) + 4 = 2 + (3 + 4)$ $(5 \times 4) \times 2 = 5 \times (4 \times 2)$ *(The group has changed but the sides are still equal.)*
Distributive	The number on the outside of the brackets is distributed to each number inside the brackets.	$12(7 + 4) = 12(7) + 12(4)$
Identity element	For multiplication, the identity element is 1. Any number multiplied or divided by 1 gives the original number. For addition, the identity element is 0.	$25 \times 1 = 25$ $3 + 0 = 3$
Zero	Any number added to 0 gives the original number. The product of any number and zero is 0.	$37 + 0 = 37$ $2000 \times 0 = 0$

Exercise 4a

1. **Calculate the following. The first one is done for you.**

   a. $17 + 32 + 96 = 145$

   b. $(17 + 32) + 96 =$

   c. $17 + (32 + 96) =$

   d. $(17 + 96) + 32 =$

2. **Do you agree that $32 \times 26 = 26 \times 32$? Explain your answer. Can you say the same for the following?**

   a. $32 \div 24$ and $24 \div 32$ b. $32 + 24$ and $24 + 32$ c. $32 - 24$ and $24 - 32$

Exercise 4b

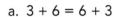

1. **Use the information above to list the properties that are represented by each of the following.**

   a. $3 + 6 = 6 + 3$

   b. $3 + 0 = 3$

   c. $4 + (6 + 2) = (4 + 6) + 2$

   d. $4 \times 1 = 4$

   e. $3 + (-3) = 0$

   f. $4(3 + 5) = 4(3) + 4(5)$

   g. $6 \times 8 = 8 \times 6$

2. **Complete the following statements.**

   a. $(8 + 4) \times 3 = (8 \boxed{} 3) + (4 \boxed{} 3)$

   b. $121 \times \boxed{} = 121$

   c. $302 \times 12 = 12 \times \boxed{}$

   d. $5697 + \boxed{} = 5697$

   e. $16 \boxed{} 0 = 16$

   f. $36 + \boxed{} = 73 + 36$

Now, try it out

Follow the clue in the first row of the table to complete the empty cells. The first one is done for you.

Determine the numbers in the empty boxes.

A	B	A + B	A × B
65	66	131	4290
34		59	
	72		504

(E) Order of operations

Points to remember

- The *order of operation* is important when solving problems involving more than one operation.

  A special set of rules can be used, known as **BODMAS** (**B**rackets, p**O**wers, **D**ivision, **M**ultiplication, **A**ddition, **S**ubtraction):

  Complete operations inside brackets first.

  Then work out all powers (squares and square roots).

  Do all multiplications and divisions next, going from left to right.

  Do all additions and subtractions last, going from left to right.

Example 1

An average of 3000 people attended each day of the Interschool Athletic Championship last year. Tickets were $2 and $1 each. If the games lasted for two days and 500 tickets were sold at $1 each, how much money was taken altogether?

Solution

The operations involved in this problem are (×, −, +).

3000 people attended the games each day for 2 days:

Total attendance	3000 × 2
	= 6000 people
500 people paying $1 per ticket	$500
Remaining attendees	3000 − 500 = 2500
2500 people paying $2 per ticket	2500 × 2 = $5000
Amount of money taken in altogether	$5000 + $500 = $5500

 Example 2

Simplify 3(16 − 5) + 8

Solution

$$3 \times (16 - 5) + 8 \quad = 3 \times 11 + 8$$
$$= 33 + 8$$
$$= 41$$

Exercise 5a

1. Simplify:

a. $8 - 4 \div 2 + 10 \div 2$

b. $3(4 + 2 - 3)$

c. $6 + 4 \times 3^2$

d. $6(12 + 8) \div 2 + 1$

e. $20 \div 4 \times 4$

2. a. Write two subtraction problems that you can solve if you know
      1286 + 4214 = 5500.

   b. Write a multiplication problem you can solve if you know
      126 ÷ 9 = 14.

   c. Write a division problem you can solve if you know
      48 × 132 = 6336.

3. A supermarket's cash sales were $5800 on Monday, $7732 on
   Tuesday and $8950 on Wednesday. If $24 482 was the total cash
   sales for the four days, how much did it sell on the fourth day?

Exercise 5b

1. Calculate:

   a. 17 − (8 + 3) =

   b. 74 − (9 + 7) ÷ 4 =

   c. 48 ÷ 6 + 6 =

   d. 74 − (8 × 7) ÷ 4 =

   e. 140 ÷ (10 × 7) =

2. Use the following numbers to complete these multiplications:
   6, 7, 8, 48, 68 and 82.

   a. ☐ × ☐ = 408   b. ☐ × ☐ = 336   c. ☐ × ☐ = 656

Now, try it out

1. **Use each of the numbers 1, 3, 5, 7 and 9 once, with any operation
   signs (+, −, ×, ÷) and grouping you wish.**

   a. Write an expression for the smallest whole number possible.

   b. Write an expression for the largest whole number possible.

PRACTICE TEST 2 – OPERATIONS ON NUMBERS
- Let's see how much you know

Section A

Select the letter of the correct answer to each question.

Two schools are playing a netball tournament. The results of the games are as follows. Use the table to answer the questions.

Game	School 1	School 2
1st	32	4
2nd	40	5
3rd	80	10
4th	96	12

1. What could be done to the score of school 1 in the 1st game to get the corresponding score for school 2?
   [a.] add 22   [b.] take away 28   [c.] divide by 8   [d.] multiply by 8   Ⓐ Ⓑ Ⓒ Ⓓ

2. What could be done to the score of school 2 in the 3rd game to make it half that of school 1?
   [a.] add 30   [b.] take away 60   [c.] divide by 4   [d.] multiply by 8   Ⓐ Ⓑ Ⓒ Ⓓ

3. What could be done to the score of school 1 in the 4th game to make it 7 times the score of school 2?
   [a.] add 30   [b.] take away 60   [c.] divide by 4   [d.] subtract 12   Ⓐ Ⓑ Ⓒ Ⓓ

4. Which equation shows the commutative property of addition?
   [a.] $5 + 1 = 1 + 5$ [b.] $3(1 + 9) = (3 + 1)9$ [c.] $2 + 7 + 5 = 14$ [d.] $0 + 9 = 9$ Ⓐ Ⓑ Ⓒ Ⓓ

5. If $8 \times 36 = 288$, what is $80 \times 36 = ?$
   [a.] 2880   [b.] 2088   [c.] 2808   [d.] 2888   Ⓐ Ⓑ Ⓒ Ⓓ

6. The difference between 6291 and n is 5524. What is n?
   [a.] 815   [b.] 762   [c.] 767   [d.] 667   Ⓐ Ⓑ Ⓒ Ⓓ

7. The answer when numbers are multiplied is called the _____.
   [a.] sum   [b.] dividend   [c.] product   [d.] operation   Ⓐ Ⓑ Ⓒ Ⓓ

8. When whole numbers are multiplied by 10 or powers of 10, which of these numbers must be in the answer?
   [a.] 1   [b.] 2   [c.] 5   [d.] 0   Ⓐ Ⓑ Ⓒ Ⓓ

9. $25\,000 \div n = 25$, so $n =$
   [a.] 10   [b.] 100   [c.] 1000   [d.] 10 000   Ⓐ Ⓑ Ⓒ Ⓓ

10. Which of these statements is correct?
    [a.] There is no relationship between multiplication and division.
    [b.] Division is repeated addition.
    [c.] Division is the inverse of multiplication.
    [d.] Addition and subtraction are the same operation.
    Ⓐ Ⓑ Ⓒ Ⓓ

11. The sum of two numbers is 16. The product of the same numbers is 63. What are the two numbers?
    [a.] 10 and 6   [b.] 16 and 1   [c.] 13 and 3   [d.] 9 and 7   Ⓐ Ⓑ Ⓒ Ⓓ

12. Which equation gives the largest result?

[a.] 22 000 × 10   [b.] 220 000 × 10   [c.] 22 000 × 100   [d.] 22 000 × 1000   (A)(B)(C)(D)

13. Which of these equations will give the same answer as 13 × 99?

[a.] 1300 × 25   [b.] (13 × 100) − 13   [c.] (130 × 100) − 1   [d.] 1300 × $\frac{1}{4}$ (A)(B)(C)(D)

14. The equation 30(12 + 3), if solved using the distributive property, would be written as

[a.] (30 × 12) + (3)   [b.] (30 × 12) + (30 × 3)   [c.] (30 × 10) + (30 × 3)   [d.] (30 × 15) + (30 × 3)   (A)(B)(C)(D)

15. There were 240 students at school on Monday. On Friday the attendance was $\frac{1}{4}$ of that number. How many students were at school on Friday?

[a.] 480 students   [b.] 60 students   [c.] 244 students   [d.] 24 students   (A)(B)(C)(D)

Section B

Answer the following.

1. A baker is hired to make pies for a party catering for 12 children. He has already baked 3 dozen pies. If each child should receive 12 pies, how many more pies should be baked?

_____

2. In preparation for a street parade, the participants were arranged in 7 rows with 24 individuals in each row. On the day of the parade, the leader decided to rearrange the troop because 3 persons were absent. If there are now 5 rows, (with only those present) how many persons are in each row?

_____

3. A man pays $10 000 for a new vehicle from his savings of $55 955.00. The remaining cash must be divided among 5 siblings. How much money should each receive?

_____

4. A water company has to produce water to supply players at a basketball tournament. If the company can only produce 800 bottles per day but needs to produce 2400 bottles, how many days are needed to complete the task?

_____

5. A supermarket owner wants to build a car park on a lot which is 960 sq ft. If a vehicle is allotted 48 square feet of space, how many vehicles can the parking lot accommodate at full capacity?

_____

3 WORKING WITH FRACTIONS

The word fraction originated from a Latin word '*fractio*' meaning 'to break'. The use of fractions began from the work of Egyptian mathematicians as early as 1800 BC but the concept of numerator and denominator came from Latin writers.

In this chapter you will review and reinforce your understanding of:

- writing and naming fractions
- Identifying fractions of shapes
- equivalent fractions
- methods of changing fractions from one form to another
- comparing and ordering fractions
- performing basic operations with fractions.

What you need to know

* A fraction is part of a whole. It is written as $\frac{a}{b}$ where a and b are numbers.

$$a \longleftarrow \text{This is the } numerator.$$
$$\overline{}$$
$$b \longleftarrow \text{This is the } denominator.$$

* If the numerator is less than the denominator, the fraction is a *proper fraction*, for example, $\frac{3}{4}$.

* If the numerator is greater than the denominator, then the fraction is an *improper fraction*, for example, $\frac{4}{3}$.

* A fraction can also be a part of an object, or part of a set of objects. For example;

a) Three slices of this pizza have been sold – $\frac{3}{8}$ of the pizza has been sold.

b) Ten of these twelve mangoes are ripe – $\frac{10}{12}$ of them are ripe.

* A *mixed number* is made up of a whole number and a proper fraction, for example, $3\frac{1}{2}$.

* Fractions, like whole numbers, can be added, subtracted, multiplied and divided.

Ⓐ Identifying fractions of quantities

Points to remember

- A fraction may be a part of an object, or part of a set of objects
- A fraction represents part of a whole and therefore a fraction is less than one

 **Example**

Identify the fraction that has been shaded

Solution

$\frac{1}{2}$ of the shape has been shaded (or $\frac{3}{6}$)

Exercise 1a

1. **For each diagram identify the fraction that is shaded.**

a.

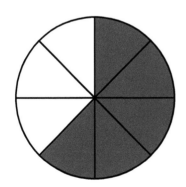

b.

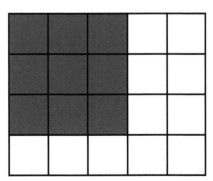

c.

d.

2. **Shade the correct fraction in each diagram.**

a. $\frac{6}{9}$ b. $\frac{2}{7}$

c. $\frac{7}{15}$ d. $\frac{8}{10}$

a.

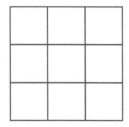

b.

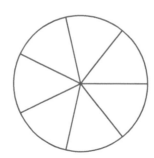

c.

d.

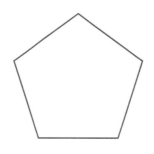

Exercise 1b

1. **Mr Brown had 30 rabbits on his farm. The table below shows the number of rabbits he sold in five days.**

Day	Number of rabbits sold
Monday	3
Tuesday	7
Wednesday	1
Thursday	5
Friday	6

a. What fraction of rabbits was sold on Tuesday?

b. What fraction was sold on Thursday?

c. What fraction was sold on Wednesday?

d. What fraction of the rabbits does Mr Brown have left?

2. Write out the following as fractions.

a. John had ten apples, he sold five. What fraction was sold?

b. Mary had eight pencils, she used up two. What fraction was left?

c. Brian had three chocolates, he ate one. What fraction did he eat?

d. Tom had 12 balloons, he popped seven. What fraction was not popped?

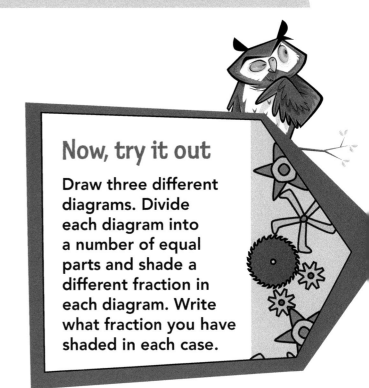

Now, try it out

Draw three different diagrams. Divide each diagram into a number of equal parts and shade a different fraction in each diagram. Write what fraction you have shaded in each case.

B Changing fractions from one form to another

Points to remember

- A mixed number can be changed to an improper fraction by multiplying the whole number by the denominator and then adding the numerator. The resulting mixed number is placed over the denominator.
- An *improper fraction* can be changed to a *mixed number* by dividing the numerator by the denominator. The resulting mixed number is made up of a whole number and a proper fraction.

Example 1

Change $\frac{72}{7}$ to a mixed number. (Part of the numerator becomes a whole number and the remainder becomes the numerator of the proper fraction.)

Solution

$\frac{72}{7} = 72 \div 7 \quad = 10 \text{ r } 2$

$\qquad\qquad = 10\frac{2}{7} \longrightarrow$ 10 is the whole number, $\frac{2}{7}$ is the proper fraction

Example 2

Change $10\frac{2}{7}$ to an improper fraction.

Solution

$10 \times 7 \longrightarrow$ whole number times denominator

$+ 2 \longrightarrow$ add numerator

$= \frac{72}{7} \longrightarrow$ place over denominator

Exercise 2a

1. **Study the diagram below.**

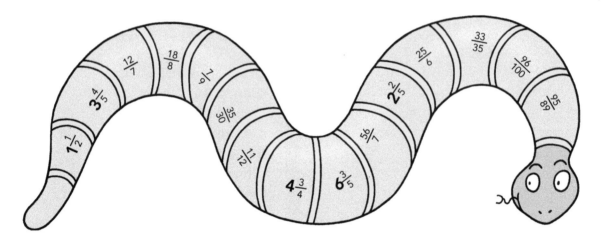

a. **Shade** the proper fractions.

b. **Circle** the improper fractions.

c. **Draw a box** around the mixed numbers.

2. **Change these improper fractions to mixed numbers.**

a. $\frac{9}{7}$ b. $\frac{29}{15}$ c. $\frac{97}{12}$ d. $\frac{166}{10}$

3. **Change these mixed numbers to improper fractions.**

a. $3\frac{1}{2}$ b. $15\frac{1}{3}$ c. $22\frac{1}{5}$ d. $18\frac{3}{4}$

Exercise 2b

1. **Complete the table below by inserting the correct answers in the empty spaces.**

Improper fraction	$\frac{8}{7}$	$\frac{15}{9}$			$\frac{19}{7}$	$\frac{35}{6}$
Mixed number equivalent	$1\frac{1}{7}$		$3\frac{1}{7}$	$4\frac{1}{8}$		

Now, try it out

a. Write down five proper fractions of your own.

b. Write down five improper fractions of your own.

c. Write down five mixed numbers of your own.

ⓒ Equivalent fractions

Points to remember

- Equivalent fractions are fractions that represent the same quantity.

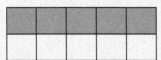

6 out of 12 are shaded = $\frac{1}{2}$

5 out of 10 are shaded = $\frac{1}{2}$

- Multiplying or dividing both the numerator and the denominator by the same number gives an equivalent fraction.

 Example

What is the missing number?

$$\frac{5}{8} = \frac{\square}{16}$$

Solution

Look at the denominators, 8 and 16.

Since 8 × 2 = 16, multiply 5 by 2, also, to get 10.

Or, cross-multiply: 8 × $\square$ = 5 × 16

$$\square = \frac{5 \times 16}{8} = 10$$

The answer is 10.

Exercise 3a

1. **Write two fractions that are equivalent to each of these.**

   a. $\frac{1}{2}$ b. $\frac{8}{12}$ c. $\frac{1}{3}$ d. $\frac{14}{35}$ e. $\frac{15}{18}$

2. **In each of the following figures shade the equivalence of the fractions provided.**

   a. $\frac{3}{5}$ b. $\frac{1}{4}$ c. $\frac{9}{12}$ d. $\frac{1}{5}$

 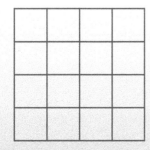  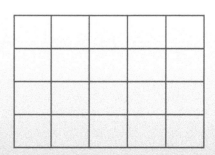

Exercise 3b

1. Place a fraction in the box which is equivalent to the one given.

a. $\frac{1}{2}$ $\boxed{\dfrac{12}{24}}$

b. $\frac{12}{16}$ $\boxed{\dfrac{}{4}}$

c. $\frac{1}{6}$ $\boxed{\dfrac{20}{}}$

d. $\frac{2}{3}$ $\boxed{\dfrac{}{9}}$

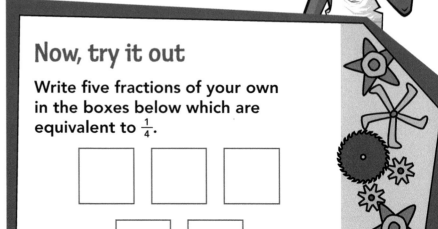

Now, try it out

Write five fractions of your own in the boxes below which are equivalent to $\frac{1}{4}$.

(D) Comparing fractions

Points to remember

- We use equivalence to compare and order fractions.
- When identifying equivalent fractions it is useful to find the lowest common denominator (LCD) of the denominators so that like denominators can be compared.
- With like denominators the size of the numerator will determine the size of the fraction.

 ### Example

Compare $\frac{7}{8}$ and $\frac{4}{5}$.

Solution

The LCD of 5 and 8 is 40

$\frac{7}{8} = \frac{35}{40}$ $\frac{4}{5} = \frac{32}{40}$

The answer is $\frac{7}{8} > \frac{4}{5}$.

Exercise 4a

1. **Rewrite each of these fractions using a common denominator and insert the appropriate sign, < or >**

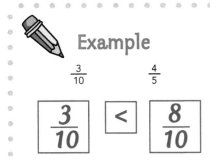

Example

$\frac{3}{10}$ $\quad$ $\frac{4}{5}$

$\boxed{\frac{3}{10}}$ $\boxed{<}$ $\boxed{\frac{8}{10}}$

a. $\frac{7}{8}$ ☐ $\frac{2}{4}$ b. $\frac{2}{3}$ ☐ $\frac{6}{8}$

c. $\frac{5}{6}$ ☐ $\frac{4}{8}$ d. $\frac{3}{9}$ ☐ $\frac{4}{10}$

2. **Write the fractions below in order of size from largest to smallest in the boxes below.**

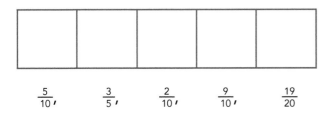

$\frac{5}{10},$ $\qquad$ $\frac{3}{5},$ $\qquad$ $\frac{2}{10},$ $\qquad$ $\frac{9}{10},$ $\qquad$ $\frac{19}{20}$

Exercise 4b

1. **Complete the boxes below to show the fractions shown on the number line.**

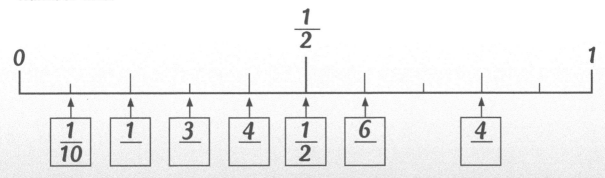

2. Complete these fraction statements by placing < or > in the boxes provided.

a. $\frac{7}{15}$ ☐ $\frac{2}{5}$ b. $\frac{3}{10}$ ☐ $\frac{4}{5}$ c. $\frac{9}{30}$ ☐ $\frac{8}{15}$ d. $\frac{3}{4}$ ☐ $\frac{11}{12}$

3. A bird carries $\frac{4}{5}$ of the material needed to build its nest while another bird carries $\frac{7}{10}$. Which bird carries the most?

4. Write in ascending order (smallest to largest):

a. $\frac{2}{5}$, $\frac{3}{4}$ and $\frac{3}{8}$

b. $\frac{10}{12}$, $\frac{5}{8}$ and $\frac{3}{4}$

c. $\frac{4}{5}$, $\frac{11}{30}$ and $\frac{5}{6}$

d. $\frac{2}{7}$, $\frac{2}{3}$ and $\frac{1}{6}$

Now, try it out

Write the fractions below on the bananas in ascending order using the same denominator.

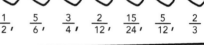

$\frac{1}{2}$, $\frac{5}{6}$, $\frac{3}{4}$, $\frac{2}{12}$, $\frac{15}{24}$, $\frac{5}{12}$, $\frac{2}{3}$

Ⓔ Reducing fractions

Points to remember

- To write a fraction in its simplest form, divide the numerator and denominator by their *highest common factor* (HCF).
- The highest common factor of two number represents the highest factor which is common to both numbers.

 Example

What is the simplest form of $\frac{10}{16}$?

Solution

$10 \div 2 = 5$ ⟶ no number except 1 will divide exactly into 5

$16 \div 2 = 8$ ⟶ and 8, so 5 and 8 are the simplest form

The answer is $\frac{5}{8}$.

Exercise 5a

1. Write these fractions in their simplest form.

   a. $\frac{5}{10}$ b. $\frac{75}{200}$ c. $\frac{12}{24}$ d. $\frac{3}{9}$ e. $\frac{9}{36}$

2. Write the fractions in ascending order on the ladder.

   $\frac{3}{10}$ $\frac{1}{5}$ $\frac{3}{8}$ $\frac{3}{4}$ $\frac{1}{4}$ $\frac{1}{2}$

Exercise 5b

1. The fraction $\frac{25}{6}$ is between which two whole numbers?

2. What fraction of 70 marbles will each boy get if the marbles are equally shared among five boys?

3. Write the smallest fraction in the box.

   $\frac{1}{9}$, $\frac{2}{8}$, $\frac{1}{2}$, $\frac{3}{4}$, $\frac{2}{3}$ $\boxed{}$

4. Write the largest fraction in the box.

   $\frac{8}{9}$, $\frac{2}{3}$, $\frac{8}{20}$, $\frac{5}{6}$, $\frac{1}{5}$ $\boxed{}$

5. A bar of chocolate has a mass of $4\frac{3}{8}$ kg. Another bar has a mass of $4\frac{9}{16}$ kg. Which bar is heavier?

Now, try it out

Match each item from Set A with the related definition in Set B.

Set A	Set B
Denominator	1. Two or more fractions that name the same amount.
Dividing	2. A number that names a part of a whole or a part of a set.
Equivalent fractions	3. One part out of three parts.
Fraction	4. Number written below the bar in a fraction.
Improper fraction	5. An operation involving sharing.
Multiple	6. Has a numerator smaller than its denominator.
A number	7. Number written above the bar in a fraction.
Numerator	8. The amount being counted or measured.
Proper fraction	9. Has a numerator larger than its denominator.
Thirds	10. A product of two numbers.

(F) Adding and subtracting fractions

Points to remember

- When adding or subtracting fractions with the same denominator, only add or subtract the numerators.
- To add or subtract fractions with different denominators, first find the *lowest common denominator* (LCD) and rewrite the fractions as equivalent fractions. Then add or subtract the numerators.
- When working with mixed numbers, change them to improper fractions before finding the lowest common denominator or find the LCD of the fractions and add or subtract the whole numbers based on the operation.

Example 1

Find the sum and difference of $\frac{2}{3}$ and $\frac{1}{6}$.

Solution

LCD = 6

$\frac{2}{3} = \frac{4}{6}$

Sum $\qquad$ Difference

$\frac{4}{6} + \frac{1}{6} = \frac{5}{6}$ $\qquad$ $\frac{4}{6} - \frac{1}{6} = \frac{3}{6} = \frac{1}{2}$ $\longrightarrow$ $\frac{3}{6}$ is simplified to $\frac{1}{2}$

Example 2

Find the sum and difference of $2\frac{1}{4}$ and $1\frac{3}{5}$.

Solution

$2\frac{1}{4} = \frac{9}{4}$ $\quad$ $1\frac{3}{5} = \frac{8}{5}$ $\longrightarrow$ LCD = 20

$\frac{9}{4} = \frac{45}{20}$ $\qquad$ $\frac{8}{5} = \frac{32}{20}$

Sum

$\frac{45}{20} + \frac{32}{20} = \frac{77}{20} = 3\frac{17}{20}$

Difference

$\frac{45}{20} - \frac{32}{20} = \frac{13}{20}$

Exercise 6a

1. **The table shows the masses of several students at school.**

Students	Riki	Oto	Naty	Joey	Dani	Sam
Mass (kg)	$52\frac{1}{4}$	$38\frac{1}{4}$	$48\frac{5}{6}$	$53\frac{9}{20}$	$51\frac{7}{10}$	$53\frac{2}{5}$

Use the table to find:

a. The student with the greatest mass.

b. The difference in the masses of Dani and Oto.

c. The combined mass of Naty and Riki.

d. The total mass of Oto, Joey and Sam.

e. Which of the students has a mass closest to 53 kg?

2. In a class of students, $\frac{5}{8}$ of the students are Scouts and $\frac{1}{3}$ are Brownies. What fraction of the class does not belong to Scouts or Brownies?

3. Two pipes are attached to a tank. After 20 minutes the larger pipe has filled $\frac{3}{5}$ of the tank and the smaller pipe has filled $\frac{1}{4}$ of the tank. How much of the tank remains to be filled?

Exercise 6b

1. A boy shared a large pineapple among his classmates during snack time. The table below shows the fraction received by each student.

Student	Fraction received
John	$\frac{1}{4}$
Mark	$\frac{1}{8}$
Javin	$\frac{3}{16}$
Tessa	$\frac{3}{8}$
Nick	$\frac{1}{16}$

   a. What fraction did Mark and John receive together?

   b. How much more did Tessa receive than John ?

   c. What fraction did Mark and Nick receive together?

   d. Who received the greatest fraction of the pineapple?

   e. How much of the pineapple was shared altogether?

2. The basketball court covers $\frac{1}{2}$ of a school yard. The netball court covers $\frac{1}{3}$ of the yard.

   a. How much more of the school yard is covered by the basketball court than by the netball court?

   b. What fraction of the school yard is covered in total by the two courts?

Now, try it out

Select a fraction and write it down on your book. Write down another fraction which is $4\frac{1}{2}$ larger than the one you selected. Write down a new fraction which is $1\frac{1}{4}$ smaller than your first fraction. Now try the same with a newly selected fraction.

(G) Multiplying fractions

Points to remember

- To multiply fractions, multiply the numerators, multiply the denominators, then write the resulting fraction in its lowest terms; or apply the operation of cancelling and then multiply the numerators and denominators.
- Always convert a mixed number to an improper fraction before multiplying.

 **Example 1**

What is $\frac{2}{3}$ of 15 marbles?

Solution

$\frac{2}{3} \times \frac{15}{1} = \frac{2 \times 15}{3 \times 1} = \frac{30}{3} = 10$ marbles ⟶ 'of' means 'multiply'

or $\frac{2 \times \overset{5}{\cancel{15}}}{\underset{1}{\cancel{3}} \times 1} = \frac{2 \times 5}{1} = 10$ marbles

 Example 2

Calculate $\frac{1}{3}$ of $\frac{6}{7}$.

Solution

$\frac{1}{3} \times \frac{6}{7} = \frac{1 \times 6}{3 \times 7} = \frac{6}{21} = \frac{2}{7}$

Exercise 7a

1. The diagram shows the position of the clothes closet in Jenny's bedroom.

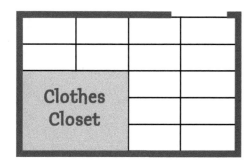

   What fraction of the room is occupied by the clothes closet?

2. In one week, Annie worked $7\frac{1}{2}$ hours each day from Monday to Friday.
   What is the total of the hours she worked that week?

3. Calculate:

   a. $\frac{5}{8} \times \frac{7}{10}$ b. $\frac{9}{16} \times \frac{4}{6}$ c. $2\frac{1}{5} \times 3$

   d. $\frac{6}{7} \times \frac{4}{7}$ e. $6\frac{2}{3} \times 4\frac{3}{8}$ f. $\frac{3}{4} \times 2\frac{1}{2} \times 1\frac{1}{2}$

Exercise 7b

1. Barry ate $\frac{1}{2}$ an orange. Brian ate $6\frac{1}{4}$ times as many. How many oranges did Brian eat?

2. A cyclist travelled $2\frac{3}{4}$ miles. A driver travelled $1\frac{1}{2}$ times that distance. How far did the car travel?

3. Complete the table below.

Fraction	Fraction multiplied by 2	Fraction multiplied by $6\frac{1}{2}$
$7\frac{1}{2}$		
$3\frac{4}{6}$		
$1\frac{3}{4}$		
$\frac{6}{8}$		

Now, try it out

Complete the table using your own fractions.

Fraction	Fraction multiplied by $\frac{1}{2}$	Fraction multiplied by $5\frac{1}{4}$

(H) Dividing fractions

Points to remember

- To divide fractions, invert (turn upside down) the second fraction and multiply.
- Remember to reduce where necessary.
- Remember to convert mixed numbers to improper fractions before dividing.
- Fractions can also be divided using the Lowest Common Denominator in which case you divide the numerators.

Example 1

$\frac{1}{2} \div 3$

Solution

$\frac{1}{2} \div \frac{3}{1}$

$= \frac{1}{2} \times \frac{1}{3} = \frac{1 \times 1}{2 \times 3} = \frac{1}{6}$ ⟶   change ÷ to × and invert second fraction from $\frac{3}{1}$ to $\frac{1}{3}$

Example 2

$\frac{3}{4} \div \frac{5}{8}$ **Solution**

$\frac{3}{4} \div \frac{5}{8}$

$= \frac{3 \times 8}{4 \times 5} = \frac{24}{20} = \frac{6}{5}$ ⟶   change ÷ to × and invert second fraction from $\frac{5}{8}$ to $\frac{8}{5}$

$= 1\frac{1}{5}$ ⟶ $1\frac{1}{5}$ is the simplest or reduced form

or using Lowest Common Denominator:

$\frac{3}{4} \div \frac{5}{8} = \frac{6}{8} \div \frac{5}{8} = 6 \div 5 = 1\frac{1}{5}$

Example 3

$1\frac{1}{2} \div \frac{7}{4}$ **Solution**

$1\frac{1}{2} \div \frac{7}{4}$

$= \frac{3}{2} \times \frac{4}{7}$ ⟶   change $1\frac{1}{2}$ to an improper fraction; change ÷ to ×; inverting $\frac{7}{4}$ gives $\frac{4}{7}$

$= \frac{12}{14} = \frac{6}{7} = 1\frac{1}{7}$ ⟶   cancel where possible

Hint: Cancel means simplify

Exercise 8a

1. **Calculate the following:**

   a. $\frac{8}{15} \div \frac{4}{5}$ b. $3\frac{2}{3} \div \frac{8}{9}$ c. $4\frac{1}{4} \div 4$ d. $5\frac{1}{4} \div 2\frac{2}{3}$

2. **What is $\frac{3}{5}$ of 100?**

3. **Karen has $7\frac{1}{2}$ cups of cooked rice. She serves each person $\frac{1}{2}$ cup. How many people can she serve?**

Exercise 8b

1. **Study the table and answer the questions.**

Car model	Speed (miles per minute)
Toyota	$2\frac{1}{2}$
Mazda	$1\frac{3}{4}$
Nissan	$1\frac{6}{8}$
Mitsubishi	$3\frac{3}{4}$
Honda	$1\frac{1}{4}$

a. How many times faster was the Toyota than the Honda?

b. How many times faster was the Mitsubishi than the Honda?

c. Which two cars were the same speed?

2. **Terry jumped $5\frac{2}{3}$ metres in a long jump competition. Bradly jumped $3\frac{1}{4}$ metres. How many times further did Terry jump?**

3. **Pam threw a javelin $12\frac{1}{2}$ metres during a sports meet. Susan threw the javelin $9\frac{3}{4}$ metres, while Florence threw the javelin $6\frac{1}{4}$ metres.**

a. How many times further was Pam's throw as compared to that of Susan?

b. Which of the three girls threw the javelin exactly two times further than another girl?

Now, try it out

Complete the table using fractions of your own.

Fraction	Fraction divided by $2\frac{2}{3}$	Fraction divided by $4\frac{3}{4}$

PRACTICE TEST 3 - FRACTIONS

- Let's see how much you know

Section A

Select the letter of the correct answer to each question.

1. What fraction is shaded in the diagram?

   [a.] $\frac{1}{4}$ [b.] $\frac{1}{2}$ [c.] $\frac{1}{3}$ [d.] $\frac{3}{4}$ Ⓐ Ⓑ Ⓒ Ⓓ

2. What fraction is shaded?

   ◯ ⬤ ◯ ⬤ ⬤ ◯ ◯ ◯ ◯

   [a.] $\frac{1}{3}$ [b.] $\frac{3}{6}$ [c.] $\frac{4}{9}$ [d.] $\frac{1}{9}$ Ⓐ Ⓑ Ⓒ Ⓓ

3. Which fraction correctly represents the following statement?

   A dog has 12 puppies. Four of the puppies are white. ☐ represents white puppies

   [a.] $\frac{12}{4}$ [b.] $\frac{1}{12}$ [c.] $\frac{1}{3}$ [d.] $\frac{1}{2}$ Ⓐ Ⓑ Ⓒ Ⓓ

4. Which fraction correctly represents the following statement?

   Brian had 20 balls, he gave away 5. The fraction of balls he gave away is ☐

   [a.] $\frac{1}{4}$ [b.] $\frac{1}{5}$ [c.] $\frac{5}{10}$ [c.] $\frac{1}{20}$ Ⓐ Ⓑ Ⓒ Ⓓ

The table below represents the number of oranges sold at a shop in 4 weeks. There were 40 oranges in the shop altogether. Study the table and answer questions five and six.

Weeks	Week 1	Week 2	Week 3	Week 4
Oranges sold	12	9	3	10

5. What fraction of the oranges was sold in week 2?

   [a.] $\frac{9}{6}$ [b.] $\frac{9}{12}$ [c.] $\frac{9}{30}$ [d.] $\frac{9}{40}$ Ⓐ Ⓑ Ⓒ Ⓓ

6. What fraction of the oranges was sold in week 4?

   [a.] $\frac{10}{12}$ [b.] $\frac{6}{30}$ [c.] $\frac{10}{24}$ [d.] $\frac{1}{4}$ Ⓐ Ⓑ Ⓒ Ⓓ

7. Which of the fractions below is equivalent to $\frac{3}{4}$?

   [a.] $\frac{6}{10}$ [b.] $\frac{15}{20}$ [c] $\frac{4}{3}$ [d.] $\frac{20}{24}$ Ⓐ Ⓑ Ⓒ Ⓓ

8. Which of the fractions below is equivalent to $\frac{1}{5}$?

   [a.] $\frac{2}{5}$ [b.] $\frac{2}{15}$ [c.] $\frac{3}{15}$ [d.] $\frac{1}{10}$ Ⓐ Ⓑ Ⓒ Ⓓ

Study the fractions listed.

$\frac{3}{4}$, $\frac{15}{20}$, $\frac{30}{40}$, $\frac{6}{8}$

9. Which fraction below is equivalent to the ones above?

   [a.] $\frac{7}{10}$ [b.] $\frac{2}{14}$ [c.] $\frac{75}{100}$ [d.] $\frac{25}{100}$ Ⓐ Ⓑ Ⓒ Ⓓ

10. Which fraction will complete the statement?

   ☐ $>$ $\frac{3}{5}$

   [a.] $\frac{3}{4}$ [b.] $\frac{1}{5}$ [c.] $\frac{2}{7}$ [d.] $\frac{5}{12}$ Ⓐ Ⓑ Ⓒ Ⓓ

11. The fractions in the table below are in descending order. Which fraction comes next?

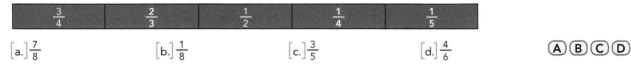

$\frac{3}{4}$	$\frac{2}{3}$	$\frac{1}{2}$	$\frac{1}{4}$	$\frac{1}{5}$

[a.] $\frac{7}{8}$ [b.] $\frac{1}{8}$ [c.] $\frac{3}{5}$ [d.] $\frac{4}{6}$ Ⓐ Ⓑ Ⓒ Ⓓ

The figure below shows the various proportions which students received from a pizza shared in class.

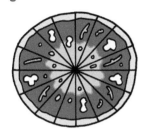

Mark $\frac{1}{8}$ Brian $\frac{7}{16}$ Randy $\frac{1}{4}$ Jerry $\frac{3}{16}$

12. How much more did Randy receive than Jerry?

[a.] $3\frac{1}{6}$ [b.] $2\frac{1}{16}$ [c.] $\frac{1}{8}$ [d.] $\frac{1}{16}$ Ⓐ Ⓑ Ⓒ Ⓓ

13. Which student received exactly $\frac{1}{8}$ more than the other?

[a.] Randy received $\frac{1}{8}$ more than Mark   [b.] Mark received $\frac{1}{8}$ more than Randy

[c.] Brian received $\frac{1}{8}$ more than Jerry   [d.] Jerry received $\frac{1}{8}$ more than Brian   Ⓐ Ⓑ Ⓒ Ⓓ

14. Four students had identical meals. At lunch break, John ate $\frac{2}{3}$ of his meal, Mary ate $\frac{1}{2}$ of hers, Tim ate $\frac{7}{8}$ of his, while Gregory ate $\frac{1}{5}$. Who ate the most?

[a.] John   [b.] Mary   [c.] Tim   [d.] Gregory   Ⓐ Ⓑ Ⓒ Ⓓ

15. Kaal picked $4\frac{3}{4}$ boxes of mangoes. He sold $1\frac{3}{8}$ boxes. How many boxes are left?

[a.] $3\frac{3}{8}$ boxes   [b.] $2\frac{3}{8}$ boxes   [c.] $2\frac{3}{4}$ boxes   [d.] $1\frac{1}{3}$ boxes   Ⓐ Ⓑ Ⓒ Ⓓ

Section B

Answer the following.

The table below shows the proportion of Carlton's week's expenditure.

Items	Clothing	Food	Bills	Savings	Entertainment
Proportion of money spent	$\frac{1}{8}$	$\frac{1}{4}$	$\frac{1}{2}$	$\frac{1}{16}$	$\frac{1}{16}$

1. What fraction was spent on bills and food together?

_____

2. What fraction was spent on food and clothing together?

_____

3. Calculate the following.

[a.] $1\frac{1}{2} + 3\frac{3}{4} =$ [b.] $2\frac{3}{4} - 1\frac{1}{4} =$

4. Calculate.

[a.] $\frac{1}{4} \times \frac{3}{4} =$ [b.] $\frac{1}{2} \times \frac{1}{3} =$

5. Calculate.

[a.] $\frac{1}{4} \div \frac{1}{2} =$ [b.] $\frac{5}{6} \div \frac{1}{4} =$

4 DECIMALS

The decimal system is often called base ten since it has ten as its base. It originated from the Hindu-Arabic numerals as early as 200 BC.

In this chapter you will review and reinforce your understanding of:

- reading and writing decimals
- place value of decimals
- adding, subtracting, multiplying and dividing decimals
- solving problems with decimals.

What you need to know

* Numbers can be written in decimal form by using a symbol called a *decimal point*.

(A) Reading and writing decimals

Points to remember

- When writing decimals, all numbers to the left of the decimal point are whole numbers.
- All numbers to the right of the decimal point are decimal fractions.
- A whole number can be written as a decimal by placing a decimal point.

Example

The numbers 69.99, 21.256 and 460.856 are in decimal form. How should these numbers be read?

Solution

H	T	O	.	Tenths	Hundredths	Thousandths
	6	9	.	9	9	
	2	1	.	2	5	6
4	6	0	.	8	5	6

69.99 is read as 'sixty-nine point nine nine'.
21.256 is read as 'twenty-one point two five six'.
460.856 is read as 'four hundred and sixty point eight five six'.

Exercise 1a

1. **Write the following decimals in words:**

   a. 0.023   b. 6.78   c. 5.023   d. 12.98   e. 15.304

Exercise 1b

1. **Write the following using figures:**

   a. one point five

   b. eight point six eight

   c. zero point zero zero three

   d. ten point zero three four

   e. eleven point one

Now, try it out

Write five decimals of your own in words and figures. Ensure that they are all less than 50.

(B) Comparing and ordering decimals

Points to remember

- To compare two or more decimals, ensure that they have the same number of decimal places. Then compare these numbers.
- The first place to the right of the decimal point is tenths
- The second place to the right of the decimal point is hundredths
- The third place to the right of the decimal point is thousandths

 Example 1

Write > or < between 0.746 and 0.74

Solution

0.746	has 3 decimal places
0.74	has 2 decimal places
0.74 is 0.740	→ convert 0.74 to 3 decimal places
0.746 and 0.740	→ both are now to 3 decimal places: 6 > 0

0.746 > 0.74

 Example 2

Which is larger 1.62 or 1.062?

Since both numbers have identical values in the ones column, compare the digits in the next column (tenths). The decimal 1.62 has a larger digit in the tenth column than 1.062. Therefore 1.62 is larger than 1.062.

Exercise 2a

1. For each of the sets listed, shade the largest decimal.

a. 0.8   0.78   0.83   0.87   0.783

b. 1.34   1.43   1.4   1.3   1.343

c. 5.9   5.67   5.09   5.91   5.679

d. 7.5   7.51   7.05   7   7.005

2. Write the decimals in order, starting with the smallest.

a. 5.4   5.43   4.53   4.40   5.34

b.  8.98   9.8   8.098    8.10   9.105

c.  11.34   11.343   11.45    11.098   11.5

d.  0.023   0.230    0.045   0.17   0.019

Exercise 2b

1. Write a suitable decimal which comes between the given pairs in each case. The first one is done for you.

a. 1.48 | 1.49 | 1.50

b. 7.34 | | 7.36

c. 0.7 | | 0.8

d. 4.56 | | 5.3

e. 12.03 | | 12.9

2. **Insert >, < or = between each pair of decimals to make each statement true.**

a. 0.5 ☐ 0.28

b. 8.3 ☐ 8.14

c. 2.43 ☐ 2.426

d. 6.74 ☐ 6.704

Now, try it out

Write down ten decimals of your own. Ensure that each decimal is smaller than 12. Now place the decimals in order of size, starting with the largest.

ⓒ Converting decimals to fractions

 Points to remember

To use a place value chart:

- The first digit to the right of the decimal point is in the tenths place.
- The second digit to the right of the decimal point is in the hundredths place.

 Example 1

Convert 0.8 to a fraction.

= 8 tenths = $\frac{8}{10}$ or $\frac{4}{5}$ ⟶ $\frac{4}{5}$ is the simplest form

Example 2

Convert 2.45 to a fraction.

O Tenths Hundreths

2 • 4   5   = $2\frac{45}{100}$ ⟶ 2 is the whole number, 0.45 is the decimal fraction; the last digit tells us the denominator of the fraction

= $2\frac{9}{10}$ ⟶ is the simplest form

Exercise 3a

1. **Convert to fractions:**

   a. 1.5   b. 0.08   c. 15.056

2. a. **What is the value of the '2' in 403.206?**
   b. **Write the number in words.**
   c. **Write the number as a fraction.**

Exercise 3b

1. **Complete the following table by inserting the correct decimal or fraction in the empty spaces.**

Decimal	Fraction
12.67	
	$9\frac{1}{10}$
5.004	
6.302	
	$3\frac{1}{2}$

2. **Match each decimal in the left column to the equivalent fraction on the right.**

3.045
3.73
0.50
0.27
3.092
0.75

$\frac{1}{2}$
$\frac{3}{4}$
$3\frac{9}{200}$
$3\frac{92}{1000}$
$3\frac{73}{100}$
$\frac{27}{100}$

Now, try it out

Write down ten decimals which are less than 20. Now, convert each decimal to a fraction.

D Converting fractions to decimals

Points to remember

- To write a fraction as a decimal, divide the numerator of the fraction by the denominator.

or:

- Find an equivalent fraction whose denominator is a power of 10, then convert to a decimal.

 Example

Write $\frac{3}{4}$ as a decimal.

Solution

$\frac{3}{4}$ is the same as $3 \div 4$

$$
\begin{array}{r}
0.75 \\
4\overline{)3.00} \\
\underline{28} \\
20 \\
\underline{20} \\
0
\end{array}
$$

$\frac{3}{4} = 0.75$

or

$\frac{3}{4} = \frac{75}{100} = 0.75$

0.75 is a terminating decimal.

This means that the fraction converts exactly into a decimal with no recurring digits.

Exercise 4a

1. **Convert to decimals:**

   a. $\frac{1}{6}$ b. $\frac{1}{4}$ c. $\frac{17}{12}$ d. $\frac{5}{8}$

2. **Complete the following table by inserting the corresponding fractions.**

Fraction	Decimal
	7.75
	15.25
	9.5
	12.34
	8.5

Exercise 4b

1. The class teacher of Grade 6 decided to share apples among her students. Jane received $5\frac{1}{4}$ apples. Represent the number of apples as a decimal.

2. Ten nuts were shared among four boys. Each boy received $2\frac{1}{2}$ nuts. Represent the amount which each boy received as a decimal.

3. A pizza was shared among 5 students.
   The fraction received by each student is shown below. Represent each fraction as a decimal.

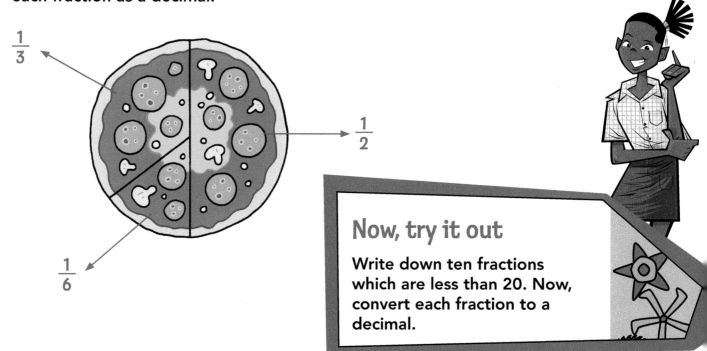

$\frac{1}{3}$

$\frac{1}{2}$

$\frac{1}{6}$

Now, try it out

Write down ten fractions which are less than 20. Now, convert each fraction to a decimal.

(E) Rounding off decimals

Points to remember

To round off decimals:

- If the digit to the right of the given place value is less than 5 the last digit remains unchanged. Drop all other digits.
- If the digit to the right of the given place value is greater than or equal to 5, the last digit is increased by 1. Drop all other digits.

Example

Round 69.37 to the nearest tenth.

Solution

In 69.37 the 3 is in the tenth place, 7 is to the right of 3 and it is more than 5.

So the 3 is increased to 4.

69.37 to the nearest tenth is 69.4.

Exercise 5a

1. **Round to the nearest tenth:**

   a. 0.94   b. 7.66   c. 48.05

2. **On average a cricketer scores 46.56 runs per match. What is this to:**

   a. the nearest tenth   b. the nearest whole number?

3. **Complete the table by writing in the correct decimals.**

Decimal	To the nearest tenth	To the nearest whole number
20.75		
15.07		
12.30		
1.06		
3.13		

Exercise 5b

1. **Jane scores 66.75 in a quiz. Write her score to:**

   a. the nearest tenth

   b. the nearest whole number.

2. **Jeremy spent $46.69 on stationery. How much did he spend:**

   a. to the nearest tenth of a dollar

   b. to the nearest dollar?

3. John runs a series of heat races. His times are all listed in the table below. Write his times to the nearest hundredth of a second and then to the nearest second.

Heat number	Time (in seconds)	Time to the nearest hundredth of a second	Time to the nearest second
1	54.567		
2	52.340		
3	49.085		
4	48.993		
5	47.047		

Now, try it out

Write a decimal number with 3 decimal places. Now, round the number to the nearest hundredth, to the nearest tenth and to the nearest whole number.

(F) Adding and subtracting decimals

Points to remember

- To add or subtract decimals, ensure that the decimal points are aligned. Fill any empty decimal places with zeros.

 ## Example

What is the sum of 51.23, 0.4 and 347?

Solution

For 'tenths', 'hundredths' and 'thousandths' we can write 'Tth', 'Hth' and 'THth'.

H	T	O	.Tth	Hth
	5	1	. 2	3
		0	. 4	0
3	4	7	. 0	0
3	9	8	. 6	3

→ align decimal points; put zero in empty decimal places

Exercise 6a

1. The following table shows the distance of students' homes from school. Study the table and answer the questions.

Student	Distance from school (metres)
Brian	41.34
Marley	91
Cuthbert	94.05
Rena	67.765
Glenda	77.078

   a. What is the combined distance of Cuthbert and Marley from the school?

   b. What is the combined distance of Rena, Brian and Marley?

   c. What is the difference between Cuthbert's distance and that of Brian?

   d. What is the difference between Glenda's distance and that of Rena?

2. Three lengths of wood are cut from a piece of wood 100.4 metres long.
   If the three lengths are 21.4 m, 19.35 m and 38.425 m, what length of wood remains?

3. How much larger is 7 than 6.482?

4. Calculate:

   a. $60.84 + 9.863 + 0.2101$ b. $60.8 - 19.735$

5. The data in the table below shows the area (in square kilometres) of several towns.

Place	Area (sq. kilometres)
Bridgetown	131.28
Castries	94.710
St Georges	9.236
Roseau	128.51
Kingston	114.23

   a. Find the total area of Bridgetown, Castries and St Georges.

   b. Find the total area of Roseau and Kingston.

   c. What is the difference between the areas of Bridgetown and Roseau?

Exercise 6b

Study the map below and answer the questions which follow.

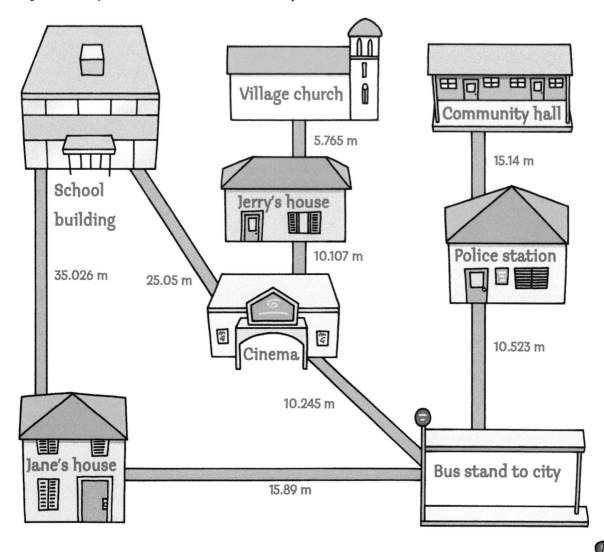

a. What is the combined distance from the bus stand to the community hall?

b. What is the combined distance from the cinema to the village church?

c. Jane walks from her house to the school building and then to the cinema. What distance did she travel in total?

d. Jerry walks from his house to the cinema and then to the bus stand. What distance did he travel in total?

e. What is the difference between the distance from Jane's house to the bus stand and the distance from the cinema to the bus stand?

Now, try it out

Write down two decimals between 56.435 and 57.278. Find the sum of the decimals which you have written.

G Multiplying and dividing decimals

Points to remember

When multiplying decimals:
- Multiply as done with whole numbers.
- Count the total number of decimal places (digits after the decimal point) in the multiplier and multiplicand.
- Put that many decimal places in the product (or answer). Care must be taken when placing the decimal point in the product.

When dividing decimals:
- Move the decimal point the same number of places in both numbers until the divisor is a whole number and then divide.

When multiplying or dividing by powers of 10:
- $16 \times 10 = 160$ (decimal point moves 1 place right of 6)
- $16 \times 100 = 1600$ (decimal point moves 2 places right of 6)
- $16 \div 10 = 1.6$ (decimal point moves 1 place left of 6)
- $16 \div 100 = 0.16$ (decimal point moves 2 places left of 6)
- $16 \div 1000 = 0.016$ (decimal point moves 3 places left of 6)
- 0 is used as a place holder.

Example 1

Six tables are lined up end to end. Each table is 2.3 m long. How long is the line of tables?

Solution

$$\begin{array}{r} 2.3 \\ \times \quad 6 \\ \hline 13.8 \end{array}$$

The answer is 13.8 m.

Example 2

4.5 × 6.3
(4.5 ⟶ one digit after the decimal point)
(6.3 ⟶ one digit after the decimal point)

Solution

$$\begin{array}{r} 45 \\ \times \quad 63 \\ \hline 135 \\ 270 \quad \\ \hline 2835 \end{array}$$

In total there are two decimal places in the numbers being multiplied. Therefore, the answer must have *2 decimal* places.

Answer: 28.35

Example 3

Calculate 7.25 ÷ 0.25

Solution

Multiply numerator and denominator by the same power of 10.
This multiplication needs to make the divisor a whole number.

$$\frac{7.25}{0.25} \times \frac{100}{100} = \frac{725}{25} = 29$$

or: $0.25\overline{)7.25} = 25\overline{)725}^{\;\;29}$

Exercise 7a

1. **A box has a height of 1.75 metres. If a number of identical boxes were all stacked up together what would be the height of:**

    a. 6 boxes   b. 9 boxes?

2. **A bus can travel 30.23 miles on a gallon of petrol.**

    a. How far can the bus go on 4 gallons of petrol?

    b. How far can the bus go on 7 gallons of petrol?

    c. How far can the bus go on 12 gallons of petrol?

3. **Cloth is sold at $2.50 a yard. Catherine bought 8.5 yards. How much did she pay?**

$2.50 per yard

4. **Twelve students run a relay race. Their average time is 35.68 seconds. What is their combined time?**

Exercise 7b

1. Oil is sold at $12.90 per litre. Sandra bought 2.7 litres of oil. How much did she pay?

2. 15 boxes have a combined height of 123 metres.

   a. What is the height of one box?

   b. If 25 boxes were stacked up, what would be the total height?

3. Calculate:

   a. 3.56 × 7.89

   b. 0.024 ÷ 1.2

   c. 0.45 ÷ 0.5

   d. 30.4 × 1.2

Now, try it out

Write down three decimals between 25.44 and 25.51

a. Multiply each of the decimals by 5.

b. Divide each of the decimals by 5.

c. Write your answers in the table below.

Decimal	Multiplied by 5	Divided by 5

PRACTICE TEST 4 - DECIMALS

- Let's see how much you know

Section A

Select the letter of the correct answer to each question.

1. **36.009 written in words is**
   - a. thirty-six point nine
   - b. thirty-six point zero zero nine
   - c. thirty-six point nine zero
   - d. thirty-six point zero nine zero

   Ⓐ Ⓑ Ⓒ Ⓓ

2. **One point zero two five written in figures is**
   - a. 1.025
   - b. 1.205
   - c. 1.005
   - d. 1.0025

   Ⓐ Ⓑ Ⓒ Ⓓ

Five students of Grade 6A participated in a relay race. Their times were recorded as follows:

Student	Time in seconds
Mary	34.098
Sandra	33.996
Christelle	35.001
Pam	35.042

3. **Which student recorded a time closest to 35?**
   - a. Mary
   - b. Sandra
   - c. Christelle
   - d. Pam

   Ⓐ Ⓑ Ⓒ Ⓓ

4. **Which student recorded a time closest to 33?**
   - a. Mary
   - b. Sandra
   - c. Christelle
   - d. Pam

   Ⓐ Ⓑ Ⓒ Ⓓ

5. **What is the combined time of Sandra and Mary?**
   - a. 68.009 s
   - b. 68.940 s
   - c. 68.094 s
   - d. 68.904 s

   Ⓐ Ⓑ Ⓒ Ⓓ

6. **What is the combined time of Christelle and Sandra?**
   - a. 67.98 s
   - b. 68.97 s
   - c. 68.907 s
   - d. 68.997 s

   Ⓐ Ⓑ Ⓒ Ⓓ

7. **What is the difference between the times of Pam and Sandra?**
   - a. 1.46 s
   - b. 1.406 s
   - c. 1.046 s
   - d. 1.004 s

   Ⓐ Ⓑ Ⓒ Ⓓ

8. **Which of these decimals come between the two given decimals?**

   11.408 ☐ 11.500

   - a. 11.400
   - b. 11.555
   - c. 11.410
   - d. 11.505

   Ⓐ Ⓑ Ⓒ Ⓓ

9. **Which of these decimals correctly completes the statement?**

   34.67 > ☐

   - a. 34.603
   - b. 34.69
   - c. 35.94
   - d. 34.70

   Ⓐ Ⓑ Ⓒ Ⓓ

10. **What is the value of the '5' in 203.256?**
    - a. 5 units
    - b. 5 tenths
    - c. 5 hundredths
    - d. 5 thousandths

    Ⓐ Ⓑ Ⓒ Ⓓ

11. **What is the value of the '4' in 504.201?**
    - a. 4 units
    - b. 4 tenths
    - c. 4 hundredths
    - d. 4 thousandths

    Ⓐ Ⓑ Ⓒ Ⓓ

12. **Susan received $\frac{2}{3}$ of an apple for her snack. Which of these decimals best represents the amount which she received?**
    - a. 2.33
    - b. 2.3
    - c. 2.66
    - d. 0.66

    Ⓐ Ⓑ Ⓒ Ⓓ

The table below represents the distance of five communities from the city. Study the table and answer the questions.

Community	Distance from the city (in miles)
Belle Mont	47.34
View Park	3.6
Soufriere	7.54
Loubiere	2.67
May Field	47.034

13. **Which community is furthest from the city?**

  a. Belle Mont   b. May Field   c. View Park   d. Loubiere   Ⓐ Ⓑ Ⓒ Ⓓ

14. **What is the difference between the distances of Belle Mont and May Field from the city?**

  a. 3.6 miles   b. 3.06 miles   c. 0.306 miles   d. 0.36 miles   Ⓐ Ⓑ Ⓒ Ⓓ

15. **A board is 3.46 inches thick. If 8 such boards are stacked up together, what would be the total thickness?**

  a. 27 inches   b. 27.06 inches   c. 27.608 inches   d. 27.68 inches   Ⓐ Ⓑ Ⓒ Ⓓ

Section B

Answer the following.

1. a. Write 23.03 in words.
    _____

    b. Write 891.762 in words.
    _____

    c. Write fifty-two point zero three six in figures.
    _____

2. **The decimal 15.75 as a fraction is**
    _____

3. **Marcia received a score of 74.67 in a placement test.**

    a. What is her score rounded to the nearest tenth?
    _____

    b. What is her score rounded to the nearest whole number?
    _____

4. **Calculate the following.**

    a. 35.78 + 3.762 =   _____
    b. 98.76 – 65.33 =   _____
    c. 234.9 – 21.5 =   _____

5. **Calculate the following.**

    a. 12.2 × 6.1 =   _____
    b. 13.4 × 4.2 =   _____
    c. 15.5 ÷ 0.5 =   _____

5 PERCENTAGES

The word percent comes from a Latin word '*per centum*' which means 100. Therefore percentage means out of a hundred.

In this chapter you will:

- explore and review the concept of percentages
- calculate profit and loss on various products
- solve problems involving percentages.

What you need to know

- ✱ Percentage means per hundred. For example 15% means 15 out of 100 or $\frac{15}{100}$.
- ✱ To find a percentage of a number, change the percentage to a fraction or decimal and then multiply by the number.

(A) Representing fractions and decimals as percentages

Points to remember

- A fraction with a denominator of 100 is called a *percentage*.
- When changing fractions to percentages, multiply the fraction by $\frac{100}{1}$.
- When changing percentages to fractions, write the percentage as a fraction with a denominator of 100.

 Example 1

Write 0.75 as a percentage.

Solution

$0.75 \times 100 = 75\%$

 Example 2

Write $\frac{7}{8}$ as a percentage.

Solution

$\frac{7}{8} \times \frac{100}{1} = \frac{700}{8} = \frac{175}{2} = 87\frac{1}{2}\%$

 Example 3

Write 42.65% as a decimal.

Solution

$42.65\% = \frac{42.65}{100} = 0.4265$

Exercise 1a

1. Write 0.75 as a percentage.

2. Write 0.40 as a percentage.

3. John received $\frac{7}{10}$ of a pizza for dinner. What percentage of the pizza did he receive?

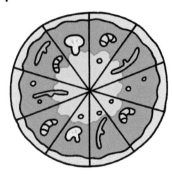

4. Susie got $\frac{35}{100}$ in a Mathematics test. Represent her score as a decimal.

5. 7 eggs out of 10 got broken. What percentage was broken?

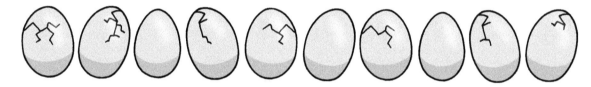

Exercise 1b

1. **Complete the table.**

Decimal	Fraction	Percentage
		5%
	$\frac{1}{4}$	
		10%
0.33		
	$\frac{2}{3}$	
		75%
0.2		
	$\frac{1}{8}$	
1.25		

Now, try it out

Write down five fractions in the table below. Convert each fraction to a decimal. Now, change each of the decimals to percentages.

Fractions	Decimal equivalent	Percentage

ⒷFinding the percentage of a number

Points to remember

■ To find a percentage of a number, change the percentage to a fraction or decimal and then multiply by the number.

 Example 1

5% of 300 =

Solution

$\frac{5}{100} \times 300 = 0.05 \times 300 = 15$

 Example 2

25% of a bag of apples are rotten. There are 14 rotten apples.
How many apples are in the bag?

Solution

25% represents 14 apples

1% represents $\frac{14}{25}$

100% represents $\frac{14}{25} \times 100 = 56$ apples

Exercise 2a

1. **Work out the following.**

   a. 25% of 250   b. 60% of 300   c. 15% of 500   d. 2% of 800

2. In a class of 60 students, 36 are girls. What percentage are boys?

3. A man sold 80 bunches of bananas at the market place. If he had 100 bunches of bananas altogether, what percentage did he sell?

4. There are 30 students in a class. 12 are absent on a given day. What percentage of the students are present?

Exercise 2b

1. 20% of the students in Grade 6 are absent. 24 students are absent. How many students are enrolled in Grade 6?

2. a. What is 25% of 20?
   b. What is 27% of 30?

3. There are 60 people at the movies. 36 of them are males. What percentage are females?

4. There are 50 cars in a car show. The table below gives the colours of the cars. Complete the table to show the fraction and the percentage of the cars represented by each colour.

Car colours	Number of cars	Fraction	Percentage
Green	5		
Red	10		
Yellow	15		
Black	12		
White	8		

Now, try it out

Write down five numbers between 50 and 75. Calculate 10% of each of your numbers.

Ⓒ Calculating the percentage change

Points to remember

To find the percentage change (increase/decrease):

$\frac{\text{Change}}{\text{Starting value}} \times \frac{100}{1}$ = Percentage change (include the % sign in the answer)

Example

If 250 is increased to 350, what is the percentage increase?

Solution

$\frac{\text{change/increase}}{250} \times \frac{100}{1} = \frac{100}{250} \times 100 = 40\%$

Exercise 3a

1. Mary's salary was increased by $50. If she previously worked for a salary of $250, what is the percentage increase?

2. 450 is decreased to 150. What is the percentage decrease?

3. At a cricket match Brian scored 130 runs in the first innings and 100 runs in the second innings. What was the percentage decrease in his score?

4. Complete the table to show the difference and the percentage decrease in each case.

Original quantity	New quantity	Difference	Percentage decrease
50 cars	30 cars		
100 students	80 students		
50 birds	40 birds		
20 books	15 books		
5 pens	3 pens		

Exercise 3b

1. There were 120 cars in a car show. At the end of the day 72 were sold. What percentage of the cars remained?

2. A student got 68 of 80 items correct in his first attempt at a test. In his second attempt he got 72 of 80 correct.

   a. What percentage did he get correct in the first attempt?

   b. What percentage did he get correct in the second attempt?

   c. What was the difference between the two attempts?

3. Complete the table to show the difference and the percentage increase in each case.

Original quantity	New quantity	Difference	Percentage increase
200	250		
125	150		
70	90		
15	25		
4	6		

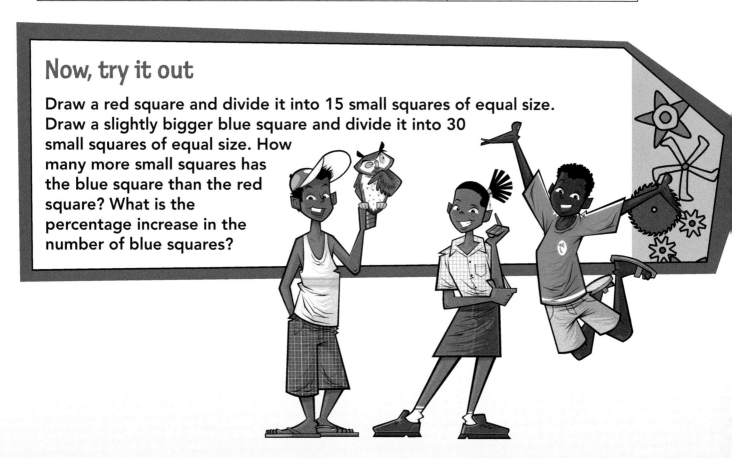

Now, try it out

Draw a red square and divide it into 15 small squares of equal size. Draw a slightly bigger blue square and divide it into 30 small squares of equal size. How many more small squares has the blue square than the red square? What is the percentage increase in the number of blue squares?

PRACTICE TEST 5 - PERCENTAGES

- Let's see how much you know

Section A

Select the letter of the correct answer to each question.

1. **Written as a percentage 0.42 is**
   - [a.] 0.42%
   - [b.] 420%
   - [c.] 4.2%
   - [d.] 42%

   Ⓐ Ⓑ Ⓒ Ⓓ

2. **Written as a percentage $\frac{32}{100}$ is**
   - [a.] 0.32%
   - [b.] 3.2%
   - [c.] 32%
   - [d.] 320%

   Ⓐ Ⓑ Ⓒ Ⓓ

3. **Twelve glasses out of a total of twenty were empty. What percentage of the glasses was full?**

   - [a.] 60%
   - [b.] 50%
   - [c.] 40%
   - [d.] 30%

   Ⓐ Ⓑ Ⓒ Ⓓ

4. **Amos received $\frac{62}{100}$ in a Science quiz. As a decimal his score is**
   - [a.] 0.6
   - [b.] 62
   - [c.] 0.62
   - [d.] 62.0

   Ⓐ Ⓑ Ⓒ Ⓓ

5. **20% of 200 boys wore red shirts. How many boys wore red shirts?**
   - [a.] 40
   - [b.] 20
   - [c.] 4
   - [d.] 2

   Ⓐ Ⓑ Ⓒ Ⓓ

6. **In a theatre there are 250 people. 100 are women. What percentage are men?**
   - [a.] 25%
   - [b.] 40%
   - [c.] 45%
   - [d.] 60%

   Ⓐ Ⓑ Ⓒ Ⓓ

7. **Five apples out of twenty went bad. What percentage went bad?**

   - [a.] 25%
   - [b.] 20%
   - [c.] 15%
   - [d.] 5%

   Ⓐ Ⓑ Ⓒ Ⓓ

8. **$\frac{12}{20}$ of a country's population is vaccinated. What percentage is vaccinated?**
   - [a.] 12%
   - [b.] 20%
   - [c.] 40%
   - [d.] 60%

   Ⓐ Ⓑ Ⓒ Ⓓ

9. **What percentage is not vaccinated?**
   - [a.] 12%
   - [b.] 20%
   - [c.] 40%
   - [d.] 60%

   Ⓐ Ⓑ Ⓒ Ⓓ

10. **There are 12 sweets in a box. Susan bought 4 boxes of sweets. What percentage of the total number of sweets can each box hold?**
    - [a.] 12%
    - [b.] 25%
    - [c.] 18%
    - [d.] 48%

    Ⓐ Ⓑ Ⓒ Ⓓ

11. **If 40% of a number is 20, what would be the number?**
    - [a.] 10
    - [b.] 40
    - [c.] 25
    - [d.] 50

    Ⓐ Ⓑ Ⓒ Ⓓ

12. **If 25 items represent 100%, which of the following would represent 10%?**
    - [a.] 2.5 items
    - [b.] 5 items
    - [c.] 100 items
    - [d.] 250 items

    Ⓐ Ⓑ Ⓒ Ⓓ

There are 50 balloons on a pole. The table below shows the various colours. Study the table and answer the questions.

Balloon colour	Number of balloons
Green	2
Red	10
Yellow	5
Black	13
White	20

13. **What percentage of the balloons is red?**

[a.] 20% [b.] 10% [c.] 5% [d.] 1% Ⓐ Ⓑ Ⓒ Ⓓ

14. **What percentage of the balloons is black?**

[a.] 25% [b.] 26% [c.] 30% [d.] 80% Ⓐ Ⓑ Ⓒ Ⓓ

15. **What would be the percentage of yellow and red balloons together?**

[a.] 30% [b.] 15% [c.] 10% [d.] 5% Ⓐ Ⓑ Ⓒ Ⓓ

Section B

Answer the following.

1. John received $\frac{20}{25}$ in a Maths quiz. What was his percentage score?

_____

2. 10% of the cows on a farm are black. 250 cows are black. How many cows are on the farm?

_____

The following table represents a company's car sales during the month of January 2010. Study the table and answer the questions which follow.

Make of vehicle	Number sold
Toyota	12
Mazda	8
BMW	3
Nissan	5
Honda	9
Mitsubishi	6
Others	7

3. **What percentage of the total sales was Mazda?**

_____

4. **What percentage of the total sales were other cars?**

_____

5. **What percentage of the total sales were Toyota and Honda combined?**

_____

6 MONEY

Money has been used in commercial transactions for thousands of years. At about 1200 BC, shells became the first medium of exchange in China.

In this chapter you will:

- improve your understanding of making purchases and giving change
- calculate profit and loss on purchases
- explore the concept of cash discounts
- develop your skills in solving problems involving money.

What you need to know

- * In writing amounts of money, we place a decimal point between the dollar and the cents.
- * We use different symbols to represent dollars and cents.

(A) Calculating amounts, making purchases and giving change

Points to remember

- When adding or subtracting amounts of money which include dollars and cents, ensure that the decimal points are aligned.
- When multiplying and dividing money use the same approach as in multiplication and division of decimals, counting the number of decimal places which will be included in the final answer.
- In making purchases we may also need to consider tax payments. Several Caribbean territories have now implemented a system of Value Added Tax or VAT.

 Example 1

Mary had $23.75. She received an additional $46.32 from her dad. How much does she have now?

Solution

$23.75 + $46.32 =
$$
\begin{array}{r}
\$23.75 \\
+ \$46.32 \\
\hline
\$70.07
\end{array}
$$

Example 2

A student received $100.00 as a prize gift. She purchased a pack of pens costing $18.50 and a school bag costing $41.50. How much change did she receive?

Solution

$100.00 − ($18.50 + $41.50) = $100.00
− $ 60.00
$ 40.00

 Example 3

If there was 15% VAT on the cost of the items how much would the student pay in taxes?

Solution

$18.50 + $41.50 = $ 60.00

$15\% \text{ of } \$60.00 = \frac{15}{100} \times \frac{60}{1} = \frac{\overset{3}{\cancel{15}}}{\underset{20}{\cancel{100}}} \times \frac{\overset{3}{\cancel{60}}}{1}$
$= 9$

The student would pay $9.00 in taxes.

Exercise 1a

1. Tracey bought a microwave for her mum's birthday gift. The item cost $500.00 plus 15% VAT. How much did she pay for the microwave.

2. Calculate the following.
   a. $234.98 + $159.06
   b. $16 509.03 + $980.07
   c. $4572.99 − $247.34
   d. $8970.00 − $2761.42

3. Sandra has six bills in her piggy bank. This amounts to $500.00. If the bills are a combination of $50 dollar bills and $100.00 bills, what combination of bills could she have? (Bills could be used more than once.)

4. A man went to the bank to cash a cheque amounting to $300.00.

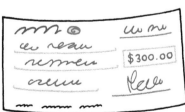

He requested $50.00 and $20.00 notes only. What possible combinations of $50.00 notes and $20.00 notes might he receive? (Remember the combinations must include $50 and $20 notes.)

Exercise 1b

1. Catherine collected $135.00 from her cake sale. Of the amount collected, she spent $73.50 purchasing school supplies and spent $14.75 on her favourite chocolates. How much does she have left?

2. Four students sold some of their favourite personal items to collect money for charity.

Josh sold a pair of trousers for $19.95

Rick sold his watch for $45.50

Kiesha sold her ring for $97.70

Brian sold his bicycle for $217.50

   a. How much did Kiesha and Brian collect for charity?

   b. How much did the students collect altogether?

3. Ice creams are sold at $3.25 for a single scoop cone and $3.75 for a double scoop cone. Susan bought 3 single scoop cones and one double scoop. How much did she pay?

Now, try it out

Use only $20.00 and $50.00 notes to make the following amounts:

a. $170.00

b. $260.00

c. $440.00

B Calculating discounts

 Points to remember
 ■ Discount means 'to take off' or 'reduce by'.

SALE! 5% off

 Example

During a Christmas sale an item costing $75 was marked down by 10%. What was the sale price?

Solution

The answer should be 10% less than the original cost.

Discount = $75 $\times \frac{10}{100}$ = $7.50 $75 – $7.50 = $67.50

or:

If 10% is taken off or discounted you will pay 100% – 10% = 90%.

Find 90% of 75.

90% = 0.90   or: $\frac{90}{\underset{4}{100}} \times \frac{\overset{3}{75}}{1}$ = 90 $\times \frac{3}{4}$ = 22.5x3 = 67.5

0.90 $\times$ 75 = $67.50

The sale price is $67.50

Exercise 2a

1. **Courts department store is having a sale with 25% off televisions.**

   a. What does '25% off' mean?

   b. If a flat screen television normally costs $4000, how much will it cost during this sale?

   c. What percentage of the original price will you be paying if you purchase the TV?

2. **A DVD player is on sale for 30% off its normal price of $2000. What is the cost of the DVD player?**

3. **A man purchased items amounting to $300 from a grocery store. He was given a discount of 5%. How much did he pay for the groceries?**

4. **Nuts are sold at 10 cents each. A discount of 5% is given for every 150 nuts purchased. How much would Kendra pay if she purchased 150 nuts?**

5. A hardware store offers 20% off on all items as part of its Mothers' Day promotion. The following represents the original price on some of the items. Work out the new prices.

$45.00

$20.00

$140.00

$450.00

$75.00

$60.00

$20.00

Exercise 2b

1. A telephone company gives a discount of 40% to customers who purchase new telephones on Mothers' Day. Dawn paid a discounted price of $200.00 for a new telephone. What was the original cost of the telephone?

2. A pair of shoes costs $120.00. The amount represents the price with a discount of 40%. What would be the actual cost of the shoes without discount?

3. The original cost of a football was $60.00. It was sold at a discounted price of $45. What was the percentage decrease in the cost of the football?

4. The shape below represents the path taken by a shopper on shopping day and the discounts which she received on items purchased. Work out the savings on each item by calculating each percentage along the shopping path.

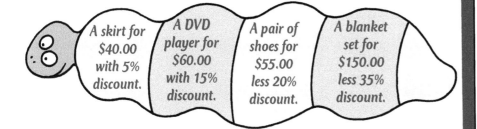

A skirt for $40.00 with 5% discount.

A DVD player for $60.00 with 15% discount.

A pair of shoes for $55.00 less 20% discount.

A blanket set for $150.00 less 35% discount.

Now, try it out

Identify six items at home or in the classroom. Estimate the cost of each of the items. Now, work out the new cost if there was a 15% discount on the estimated price for each of the items.

© Calculating profit and loss

Points to remember

- A *profit* occurs when the *selling price* (SP) is greater than the *cost price* (CP).
- A *loss* occurs when the selling price is less than the cost price.
- The percentage profit or loss can be found by using the following rules.

$$\frac{Profit}{CP} \times \frac{100}{1} \quad \text{or:} \quad \frac{Loss}{CP} \times \frac{100}{1}$$

Example

Sharon bought a pair of shoes for $65 and later sold them for $40. What was her percentage loss?

Solution

Cost price of shoes   = $65

Selling price of shoes = $40

Loss   = $65 − $40 = $25

Percentage loss   = $\frac{25}{65} \times \frac{100}{1}$ = $38.5 (1 decimal place)

Exercise 3a

1. Mr Smith bought a house for $178 000 and sold it for $200 000. What was the percentage profit?

2. The value of an airline ticket to Grenada was $400. If the fare is reduced to $320, find the percentage loss to the airline.

3. Tony bought three pens for $3.60. He later sold them at $1.50 each. Find his percentage profit.

4. Leon bought a pair of shoes for $95. If he sold them for $80, what was the loss as a percentage?

Exercise 3b

1. **Find the value of '*n*' in each of these problems.**

   a. 75% of $n = 75$ b. 40% of $n = 400$ c. 50% of $n = 440$

   d. 15% of $n = 30$ e. n% of 140 = 70

2. **A variety store offers 20% discount for every $100.00 spent and 40% discount for any additional amount spent. Joey spent $250.00 at the store.**

   a. How much discount did he receive?

   b. How much did he actually pay?

3. **A hardware store sells stoves for $1500.00 less 5% discount. The store also sells refrigerators at a cost of $3000 less 10% discount. Bradley bought a stove and a refrigerator.**

   a. How much did he have to pay altogether?

   b. How much did he save?

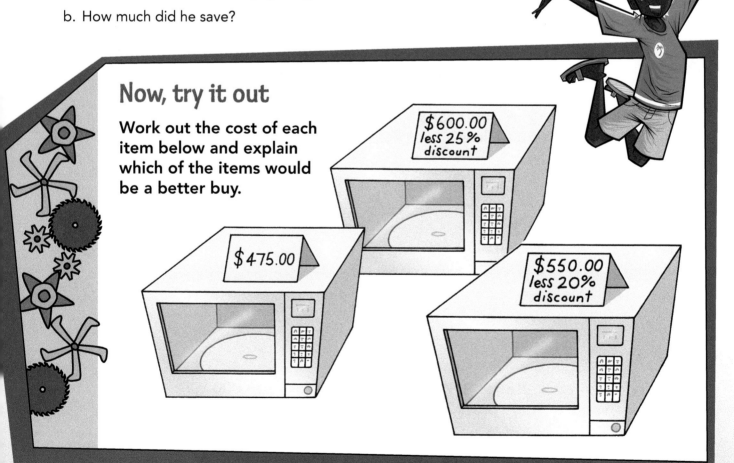

Now, try it out

Work out the cost of each item below and explain which of the items would be a better buy.

$600.00 less 25% discount

$475.00

$550.00 less 20% discount

PRACTICE TEST 6 – MONEY

- Let's see how much you know

Section A

Select the letter of the correct answer to each question.

1. The amount $297 506.01 written as words is
   - a. two hundred and ninety-seven dollars, five hundred and six dollars and one cent
   - b. two hundred and ninety-seven thousand and six dollars
   - c. two hundred and ninety-seven thousand, five hundred and six dollars and one cent
   - d. two hundred and ninety-seven thousand, five hundred, six thousand and one cent

   Ⓐ Ⓑ Ⓒ Ⓓ

2. The amount $345 897.09 written as words is
   - a. three hundred and forty-five thousand, eight hundred and ninety-seven dollars and nine cents
   - b. three hundred and forty-five thousand, eight hundred and nine cents
   - c. three hundred and forty-five thousand, eight hundred ninety-seven dollars
   - d. three hundred and fifty-eight thousand, nine hundred and seventy dollars and nine cents

   Ⓐ Ⓑ Ⓒ Ⓓ

3. Written in figures, six hundred and fifty thousand, four hundred and ninety-six dollars is
   - a. $650 496.00
   - b. $650 000 496. 00
   - c. $65 000 496.00
   - d. $650 496 000.00

   Ⓐ Ⓑ Ⓒ Ⓓ

4. Written in figures one million, six hundred thousand dollars is
   - a. $1 006 000.00
   - b. $1 006 000 000.00
   - c. $1 600 000.00
   - d. $100 600.00

   Ⓐ Ⓑ Ⓒ Ⓓ

5. Delroy deposits $150.00 on his savings account every month. The bank gives him $50.00 as a fixed interest on the total amount every year. How much would he have made in three years of constant saving?
   - a. $505.00
   - b. $5450.00
   - c. $5005.00
   - d. $5550.00

   Ⓐ Ⓑ Ⓒ Ⓓ

6. For his eleventh birthday, Jack was given $75.50 by his dad and $235.40 by his mom. How much was he given altogether?
   - a. $300.00
   - b. $3000.10
   - c. $310.90
   - d. $3010.00

   Ⓐ Ⓑ Ⓒ Ⓓ

7. Jessica had $235.50 for her weekend shopping. She bought a pair of shoes costing $35.05 and a skirt costing $120.00. How much money does she have left?

   - a. $175.05
   - b. $80.45
   - c. $270.75
   - d. $200.50

   Ⓐ Ⓑ Ⓒ Ⓓ

8. If she received 10% off on the cost of the skirt, how much would the skirt have cost?
   - a. $108.00
   - b. $110.00
   - c. $100.00
   - d. $132.00

   Ⓐ Ⓑ Ⓒ Ⓓ

9. A farmer purchases a wheelbarrow for $640.00. He receives $128.00 off. What percentage did he receive off the wheelbarrow?

   - a. 20%
   - b. 40%
   - c. 30%
   - d. 50%

   Ⓐ Ⓑ Ⓒ Ⓓ

10. A store sells a cake mixer for $350.00. It gives 50% off the cost of an iron when a cake mixer is purchased. An iron costs $140.00. Pamela purchased an iron and a cake mixer. How much would she pay?

    - a. $500.00
    - b. $490.00
    - c. $420.00
    - d. $42.00

    Ⓐ Ⓑ Ⓒ Ⓓ

11. Brenton's salary was increased by $250.00. He previously received a salary of $1000.00. What was the percentage increase?

[a.] 20%   [b.] 30%   [c.] 25%   [d.] 50%   Ⓐ Ⓑ Ⓒ Ⓓ

12. A book cost $20.00. This represented a discount of 75%. What was the original cost of the book?

[a.] $80.00   [b.] $25.00   [c.] $50.00   [d.] $20.00   Ⓐ Ⓑ Ⓒ Ⓓ

13. A flat screen television goes on sale for $4320.00. This represents a discount of 20% off the regular price. What is the regular price?

[a.] $5400.00   [b.] $5500.00   [c.] $6000.00   [d.] $6500.00   Ⓐ Ⓑ Ⓒ Ⓓ

14. A living room set has been marked down from $1000 to $500. By what percentage has the price been cut?

[a.] 25%   [b.] 50%   [c.] 75%   [d.] 80%   Ⓐ Ⓑ Ⓒ Ⓓ

15. Jack bought a skateboard for $80.00. He then sold it for $60.00. What was his percentage loss?

[a.] 10%   [b.] 25%   [c.] 20%   [d.] 30%   Ⓐ Ⓑ Ⓒ Ⓓ

Section B

Answer the following.

1. A man bought a sports car for $12 000.00. After repairs he sold it for $18 000.00. What was his percentage profit?

_____

2. Sweets costs $5.00 for a jar of 50. Jack bought 250 sweets. How much did he pay?

_____

3. A restaurant in the city charges $12.50 for a chicken lunch and $15.00 for a fish lunch. How much would 3 chicken and 4 fish lunches cost altogether?

_____

4. Six students in one Grade 6 class at the St. Martin Primary School decided to operate a joint savings account for charity. There are 30 students in the class. If each of the students saved $20.00 per month, how much would they have collected in one year?

_____

5. A man wants to purchase a bicycle with a price tag of $450.00. If he gets a 20% discount on the purchase, how much would he pay for the bicycle?

7 RATIO AND PROPORTION

Ratio is a relationship between two quantities or objects of the same kind. Ratios may have been used since pre-historic times since this method was often used in dividing quantities among populations throughout civilisation.

In this chapter you will:

- find ratios of two quantities
- calculate and compare ratios
- calculate inverse proportion.

What you need to know

* A *ratio* is a way of expressing a proportion.
* It compares two numbers or quantities.

(A) Comparing and using ratios

Points to remember

■ The symbol : is used to compare quantities as whole numbers.

 ### Example 1

A maths group is made up of 5 girls and 3 boys.

That is, for every 5 girls there are 3 boys.

The ratio of girls to boys is 5 to 3 or 5:3.

This also means $\frac{5}{8}$ of the group are girls and $\frac{3}{8}$ are boys.

(5 + 3 = 8 parts)

 ### Example 2

In Grade 6 there are 10 girls and 4 boys.

The ratio of girls to boys is 10:4 or 5:2 in its simplest form.

$\frac{10}{2}$ = 5 and $\frac{4}{2}$ = 2

Exercise 1a

1. **Study the diagram and insert the missing numbers in the following sentences.**

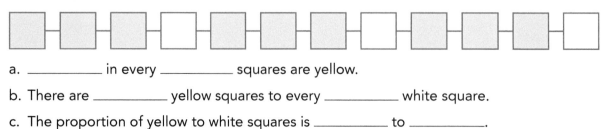

    a. _____ in every _____ squares are yellow.

    b. There are _____ yellow squares to every _____ white square.

    c. The proportion of yellow to white squares is _____ to _____.

2. **Shade the loops in the chains to match each description. Remember the chains may have different numbers of loops.**

    a. 3 in every 5 loops are red.

    b. 1 in every 7 loops is black.

    c. 2 in every 3 loops are green.

3. **At a school graduation there is 1 boy to every 4 girls. How many girls are there if there are**

    a. 4 boys?   b. 6 boys?

    c. 30 children altogether?   d. 35 children altogether?

4. **How many boys are there if there are**

    a. 8 girls?   b. 12 girls?

    c. 20 children altogether?   d. 50 children altogether?

Exercise 1b

1. **There are 30 white and black balls in a bag. How many white balls are in the bag if there are**

    a. 2 white balls to every black ball

    b. 5 white balls to every black ball

    c. 9 white balls to every black ball?

2. **How many black balls are there if there are**

   a. 5 black balls to every white ball

   b. 14 black balls to every white ball

   c. 1 black ball to every white ball?

3. **In a box there are 6 blue marbles, 9 red marbles and 12 green marbles.**

   If the marbles are shared between Sam and Margaret in the ratio of 1:2

   a. How many of each colour will Sam receive?

   b. How many of each colour will Margaret receive?

4. **A 1000 gram cake is made with 200 grams of sugar to every 500 grams of flour.**

   a. How much **flour** would be required for 600 grams of sugar?

   b. How much **sugar** would be required for a 3000 gram cake?

   c. How much **flour** would be required for a 3000 gram cake?

Now, try it out

You are given a box of 40 mangoes to share among three other children and yourself. Name the children that you will share your mangoes with. Determine what ratio you will use to share your mangoes and show the amount that each child including yourself will receive.

B Finding the ratio of two quantities

Point to remember

- To find the ratio of two quantities, both quantities must be written with the same unit of measurement. It is easier to convert the larger unit to the smaller unit.

Example

Express $4.50c as a ratio in its simplest form.

Solution

$4 = 400c so 400:50 = 8:1

Exercise 2a

1. A fruit punch is made with 50 millilitres of concentrated fruit juice to every $\frac{1}{2}$ litre of water. Express the quantities as a ratio in its simplest form.

2. A cup of coffee is made with 10 grams of sugar to 40 grams of coffee powder. How much sugar would be used for a cup with 4 grams of coffee powder?

Exercise 2b

1. Write these ratios as fractions in their simplest form.

   a. 12:16   b. 20:30   c. 25:40

   d. 6:9   e. 40:60   f. 20:100

2. A store has a promotion of 4 plates for $5.00.

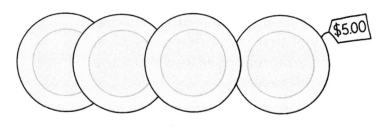

$5.00

   a. How many plates can John buy with $20.00?

   b. John has $100.00. How many plates can he buy?

   c. If John purchased 50 plates, how much would he pay?

Now, try it out

A bag of sweets contains 10 sweets. If 12 bags of sweets are shared between two children in the ratio of 1:2, how many sweets would each child get?

(c) Calculating ratios

Points to remember

- To divide something in a given ratio, work out how many parts there are altogether.
- Then find out the value of one part.

Example

Share 84 marbles between Joe and Jan in the ratio 5:7 respectively. How many does Jan receive?

Solution

Joe and Jan's shares = 5 + 7 = 12 shares

so 1 share = 84 ÷ 12 = 7

Joe gets 5 shares, Jan gets 7 shares = 7 × 7 = 49 marbles.

Exercise 3a

1. **Divide $96 among Tony, Sally and Renee in the ratio 2:1:5. What is Renee's share?**

2. a. **$90 is divided between two sisters so that the elder gets twice as much as the younger. How much does the younger sister get?**

   b. **The height of two students is in the ratio 2:3. If the shorter student's height is 120 cm, what is the height of the taller one?**

Exercise 3b

1. **The table below shows a fisherman's catch. Study the table and answer the questions.**

   a. If the fisherman shared his catch of snapper between two friends in the ratio of 2:5, how many pounds will each get?

   b. If the fisherman shares his catch of blue marlin among Pamela, Sophie and Kiesha in the ratio of 1:2:3, how many pounds would each get?

   c. If the fisherman shared his catch of tuna and red fish among his two brothers in the ratio of 1:2, how many pounds of each type would each of his brothers get?

Fish type	Amount
Red fish	24 pounds
Snapper	21 pounds
Tuna	30 pounds
Blue marlin	60 pounds

2. Pumpkin soup is ordered to serve guests at a wedding ceremony. If 1 litre of soup serves 8 people, how much soup must be ordered to serve 200 people?

Now, try it out

A 12-slice pizza is shared among three classmates. One student received 2 slices, another received 4 slices and the third received 6 slices.

In what ratio was the pizza shared?

(D) Proportion

Points to remember

- *Direct proportion.* Two quantities are in direct proportion when one increases in the same ratio as the other.
- *Inverse proportion.* Two quantities are inversely proportional if one goes up as the other goes down.

Example 1 (Direct proportion)

If 3 pens cost $3.90, what is the cost of 6 pens?

Solution

Cost of 3 pens = $3.90

Cost of 1 pen = $\frac{\$3.90}{3}$ = $1.30

Cost of 6 pens = $1.30 × 6 = $7.80

Example 2 (Inverse proportion)

If 3 men take 12 days to build a shed, how long will it take 9 men to do the same job?

Solution

3 men take 12 days

1 man will take 3 × 12 = 36 days

9 men will take $\frac{36}{9}$ = 4 days

Exercise 4a

1. 3 pens cost $6. Find the cost of 12 similar pens.

2. It took 6 students 4 days to completely mark out the football field. How long will it take 8 students to do the same job?

3. There is enough food to last 24 hikers for 30 days.

   a. If 4 hikers left camp, how long would the food last the remaining hikers?

   b. If 6 hikers join them, how long would the food last?

Exercise 4b

1. 3 woodpeckers take 4 days to bore a hole though a tree.

   a. How long would 6 woodpeckers take if they worked at the same pace?

   b. How long would 12 woodpeckers take if they worked at the same pace?

2. 25 farmers took 10 days to plant 2 acres of corn.

   a. If 50 farmers worked at the same speed, how long would it take them to plant the same amount?

   b. If 5 farmers worked at the same speed, how long would it take them to plant the same amount?

3. One batsman on the West Indies Cricket team scored 60 runs in 15 minutes.

   a. If 2 batsmen batted with the same run rate in one innings, how many runs would they make altogether in 15 minutes?

   b. If 2 batsmen batted with the same run rate in one innings, how many runs would they make altogether in 30 minutes?

Now, try it out   Body parts proportions

Using a calculator and a measuring tape, work with a partner or in a small group.

a. Measure the body parts listed below for each member of your group.

b. Round each measurement to the nearest centimetre.

c. Complete the chart.

Name	Height	Wrist	Hand span	Ratio wrist: hand span

d. What is the ratio of height to hand span for each person?

e. Who has the largest ratio of wrist to hand span?

f. Look at the ratios for the members of your group. Are any of the ratios nearly the same for all members? Explain your answer.

PRACTICE TEST - 7 RATIO AND PROPORTION

- Let's see how much you know

Section A

Select the letter of the correct answer to each question.

Study the diagram.

1. For every 4 red triangles there are _____ green triangles.

   [a.] 6   [b.] 5   [c.] 4   [d.] 2   Ⓐ Ⓑ Ⓒ Ⓓ

2. The ratio of red to green triangles is _____ to _____.

   [a.] 2:1   [b.] 2:2   [c.] 3:1   [d.] 3:2   Ⓐ Ⓑ Ⓒ Ⓓ

3. In the figure below

   [a.] 4 in every 5 shapes are blue.
   [b.] The ratio of blue to yellow shapes is 4:6
   [c.] The ratio of blue to yellow shapes is 4:2
   [d.] There are 4 blue shapes to every yellow shape.   Ⓐ Ⓑ Ⓒ Ⓓ

4. At a school graduation there are 5 girls to every 4 boys. How many girls are there if there are 16 boys?

   [a.] 20 girls   [b.] 21 girls   [c.] 30 girls   [d.] 35 girls   Ⓐ Ⓑ Ⓒ Ⓓ

5. How many boys are there if there are 30 girls?

   [a.] 20 boys   [b.] 28 boys   [c.] 24 boys   [d.] 36 boys   Ⓐ Ⓑ Ⓒ Ⓓ

6. A pack of sweets contains 28 sweets. How many green sweets are there if there are 3 red sweets to every green sweet?

   [a.] 7 green sweets   [b.] 14 green sweets   [c.] 21 green sweets   [d.] 28 green sweets   Ⓐ Ⓑ Ⓒ Ⓓ

7. What is 80 mm to 160 mm express its lowest terms?

   [a.] 8:16   [b.] 2:6   [c.] 2:4   [d.] 1:2   Ⓐ Ⓑ Ⓒ Ⓓ

8. What is 15 minutes to 1 hour expressed in its lowest terms?

   [a.] 1:4   [b.] 15:1   [c.] 15:60   [d.] 1:60   Ⓐ Ⓑ Ⓒ Ⓓ

9. A store has a promotion of 10 towels for $25.00. How many towels can Margaret buy with $75.00?

   [a.] 30 towels   [b.] 35 towels   [c.] 40 towels   [d.] 50 towels   Ⓐ Ⓑ Ⓒ Ⓓ

10. If Margaret went home with 20 towels, how much did she spend?

   [a.] $20.00   [b.] $25.00   [c.] $50.00   [d.] $75.00   Ⓐ Ⓑ Ⓒ Ⓓ

11. If 4 men took 12 days to plaster a wall, working at the same speed how long will 8 men take to complete an identical wall?

a. 10 days   b. 8 days   c. 6 days   d. 4 days   Ⓐ Ⓑ Ⓒ Ⓓ

12. If 10 students took 15 minutes to paint a wall how long would 15 students take to paint the same wall?

a. 6 minutes   b. 8 minutes   c. 10 minutes   d. 12 minutes   Ⓐ Ⓑ Ⓒ Ⓓ

13. A sum of $120.00 is shared among Jeffrey, Robert and Clement in the ratio 1:2:3. How much will Jeffrey receive?

a. $20.00   b. $40.00   c. $60.00   d. $80.00   Ⓐ Ⓑ Ⓒ Ⓓ

14. How much will Robert receive?

a. $20.00   b. $40.00   c. $60.00   d. $80.00   Ⓐ Ⓑ Ⓒ Ⓓ

15. How much will Clement receive?

a. $20.00   b. $40.00   c. $60.00   d. $80.00   Ⓐ Ⓑ Ⓒ Ⓓ

Section B

Answer the following.

1. The height of two boys is in the ratio 3:4. If the shorter student is 90 cm, what is the height of the taller one?

_____

2. There are 28 red and green sweets in a bag.

How many red sweets are there if there are 5 red sweets to every 2 green sweets?

_____

3. The distance of 300 miles from Dominica to Grenada measures 3 cm on a map. What is the scale of the map?

_____

4. $\frac{1}{2}$ litre of fruit juice comprises 100 millilitres of syrup and 400 millilitres of water. What is the ratio of syrup to water?

_____

5. How much syrup would be needed to make 1 litre of fruit juice?

_____

MEASUREMENT

Units of measure were among the earliest tools used by humans. Before units of measure were invented man used his body to make accurate measurements. Even today, we still use our body parts calculate dimensions in an accurate manner.

In this chapter you will review and extend your knowledge of:

- length, mass and temperature
- perimeter and area
- working with scale drawings
- converting units of measurement
- telling and writing time using both digital and analogue notation
- calculating the duration of events
- calculating average speed
- solving problems involving distance and time.

What you need to know

* The word 'measure' means 'a unit of comparison' we measure to find the size, quantity, amount, time or degree of.

* *Kilo* means 1000; *centi* means 100; *milli* means one thousandth.

(A) Units of measurement: length

Points to remember

Metric unit of length	Abbreviation
kilometre	km
metre	m
centimetre	cm
millimetre	mm

Length		
	10 mm	= 1 cm
	100 cm	= 1 m
	1 m	= (10 × 100) mm
	1000 m	= 1 km
	1 km	= (100 × 1000) cm

Of these four metric units of length, the *kilometre* is the longest and the *millimetre* the shortest.

■ To convert from a larger unit to a smaller unit, we multiply.

■ To convert from a smaller unit to a larger unit, we divide.

Example 1

The thickness of a 25 cent coin is best measured in millimetres.
The length of your index finger is best measured in centimetres.
The distance between countries is best measured in kilometres.

 Example 2

Write 5.3 metres in centimetres.

Solution

5.3 × 100
= 530 cm

Example 3

Write 1245 millimetres in centimetres.

Solution

1245 ÷ 10
= 124.5 cm

Exercise 1a

1. A seamstress purchased 5 metres of fabric to make wedding dresses. If she has already used up 460cm of material, how much material does she have left?

2. From John's house to Thomas' house is 10cm measured in a straight line. The houses are shown on the map below. Using the scale 1 cm = 1.5 km, what is the actual distance between the two houses?

John's house

Scale: 1cm = 1.5 km   **X**   Thomas' house

Exercise 1b

1. Measure the lines below. If 1 cm represents 405 metres, what distance does each line represent in kilometres?

   a. ————————————————————————

   b. ————————————————

   c. ——————————

2. Choose the best unit for measuring:

   a. the length of a cricket bat   b. the length of a kitchen door

   c. the distance from Dominica to St. Lucia   d. the length of your pen

   e. the length of a fly's wing

3. The distance from home to school is 16 km. I walked for 1250 metres and then took the bus for the rest of the way. How far did I travel on the bus in:

   a. metres   b. kilometres?

Now, try it out

1. What is the length of these objects to the nearest inch and cm?

Object	cm
Your exercise book	
Your longest finger	
A stick of chalk	
Your desk	

2. Measure your math textbook using a centimetre ruler. By how much is its length greater than its width?

B) Units of measurement: mass

Points to remember

- *Mass* is the amount of matter in an object.
- The *weight* of an object is the force of gravity acting on it.

Metric unit for mass	tonne	kilogram	gram	milligram
Abbreviation	t	kg	g	mg

Mass	
	1 g = 1000 mg
	1 kg = 1000 g
	1 t = 1000 kg

Example

A *milligram* is a small unit of mass.
A small photograph has a mass of about 1 *gram*.
A pineapple has a mass of about 1 *kilogram*.
The *tonne* is used to measure very heavy objects such as a car.

Exercise 2a

1. If 96 kg of cement is put into 12 kg bags, how many bags will be filled?

2. Sandra buys 14 kg of sweets for her store. She places them into bags each containing 1000 grams. How many bags will she obtain?

3. A truck weighs 3.18 tonnes. When carrying a full load the truck weighs 4.72 tonnes. What is the mass of the load in the truck?

Exercise 2b

1. Look at the large pointers on each of the scales and write the mass shown.

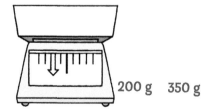   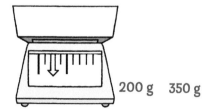

2. Choose the best unit for measuring the mass of the following items.

   a. your maths textbook   b. a truck   c. a puppy

   d. a box of chalk   e. a handkerchief

3. Flour is sold in bags of 10.8 kg. What is the weight in grams of a bag of flour?

Now, try it out

Select any five small objects which can be measured in kilograms. Estimate their masses. Find the actual mass of each object and then calculate the difference.

Object	Estimated mass	Actual mass	Difference between estimate and actual mass

Ⓒ Units of measurement: temperature

Points to remember

- A *thermometer* is used for measuring *temperature*.
- Temperature is measured in *degrees Celsius* (°C) and *degrees Fahrenheit* (°F).
- Water freezes at 0°C and boils at 100°C.
- Water freezes at 32°F and boils at 212°F.
- To convert Celsius temperatures to Fahrenheit: multiply the Celsius temperature by $\frac{9}{5}$, then add 32° to adjust for the offset in the Fahrenheit scale.
- To convert Fahrenheit to Celsius: subtract 32° to adjust for the offset in the Fahrenheit scale, then multiply the result by $\frac{5}{9}$.

Example 1

Convert 37°C to Fahrenheit.

Solution

$37 \times \frac{9}{5} = \frac{333}{5} = 66.6$

$66.6 + 32 = 98.6°F$

So 37°C = 98.6°F

Example 2

Convert 98.6°F to Celsius.

Solution

$98.6 - 32 = 66.6$

$66.6 \times \frac{5}{9} = \frac{333}{9} = 37°C$

Exercise 3a

1. **Write the temperature in degrees Fahrenheit and degrees Celsius.**

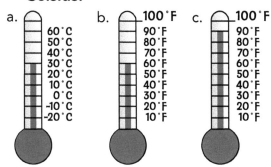

2. **Use the table to answer the questions.**

Days	Daytime temperature	Night-time temperature
Monday	28°C	22°C
Tuesday	30°C	20°C
Wednesday	34°C	24°C
Thursday	26°C	18°C
Friday	24°C	16°C

a. Which was the hottest day?

b. Which was the coldest night?

c. What was the difference between the daytime temperature and the night-time temperature on Thursday?

d. By how many degrees was Monday afternoon hotter than Friday night?

e. What was the difference between the night-time temperature on Wednesday and that on Friday?

Exercise 3b

1. The recorded temperature inside an aircraft is 12°C. The air outside is 2°C. What is the difference between the temperature in the aircraft and the temperature outside?

2. The temperature of a cup of hot coffee is 84°C. It was left standing for ten minutes reducing the temperature by 15 degrees. What is the new temperature of the cup of coffee?

3. The temperature of a glass of water is 12°C. It was placed in a freezer to become chilled. The temperature of the water was reduced by 2°C for every minute the water remained in the freezer. What would be the new temperature of the water after being in the freezer for 10 minutes?

Now, try it out

You will need a few small thermometers. Work in small groups of about four students per group.

a. Record the temperature of a glass of water at room temperature.
b. Record the temperature of a glass of slightly warm water.
c. Now record the temperature of a glass of ice cold water
d. What is the difference in temperatures in each case?
e. Record the information in the form of a table.

(D) Calculating perimeter

Points to remember

- The *perimeter* of a polygon is the distance around it.

Exercise 4a

1. An athlete runs around a tennis court that measures 96 m by 70 m. How far does he run if he goes around the court twice?

2. A rug is used to cover a floor with sides 15 metres x 20 metres. How many square metres of rug would be needed to cover the surface of the floor?

Exercise 4b

1. Find the perimeter of a kitchen floor with the plan shown (not drawn to scale).

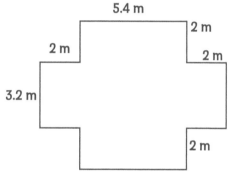

2. Find the perimeter of the following shapes.

a.
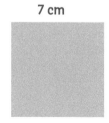
7 cm

b.
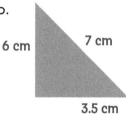
6 cm   7 cm
3.5 cm

c.

3 cm
9 cm

d.
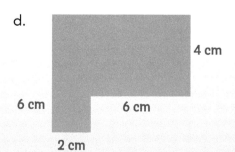
4 cm
6 cm   6 cm
2 cm

e.
3 cm
6 cm
5 cm

Now, try it out

Identify any small rectangular object in your classroom.

a. Estimate its length and width.
b. Now use your ruler to measure the actual length and width of the object.
c. Estimate the area and perimeter of the object.
d. Now calculate the area and then the perimeter using the actual dimensions.
e. What is the difference in each case?
f. Record your results in the table below.

	Estimated	Actual	Difference
Length			
Width			
Perimeter			
Area			

(E) Calculating area

Points to remember

- *Area* is the number of square units needed to cover a surface.

- Area of a square = side × side
- Area of a rectangle = length × width
- Area of a triangle = $\frac{1}{2}$ × base × vertical height
- Estimation may be used to find the area of irregular shapes.

Example

Estimate the area of this shaded shape.

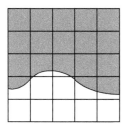

Solution

Begin by counting the number of whole squares = 13

Then the number of part squares (roughly halves) = 5

Area of shape = 13 whole sq units $+\frac{5}{2}$ sq units = 13 + 2.5 = 15.5 sq units

Exercise 5a

1. Bill is covering a table with glass. How much glass will be needed if the table measures 30 centimetres on each side?

2. The area of a square is 64 cm². What is the length of a side?

3. Jerry charges $1.00 per square metre to tile a floor. How much will he charge to tile a rectangular floor 15 m by 11 m?

4. Find the area of a square whose sides measure:

    a. 5 cm   b. 12 m

Exercise 5b

1. For each of the following shapes, find the area of the coloured triangle and then the area of the entire shape.

Figure	Area of coloured triangle	Area of shape
a.		
b.		
c.		

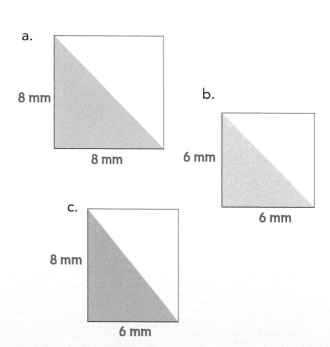

a.

8 mm

8 mm

b.

6 mm

6 mm

c.

8 mm

6 mm

2. Estimate the area of each shape and then calculate the actual area.

a.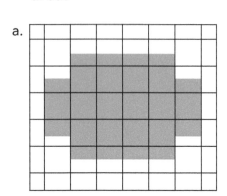

b.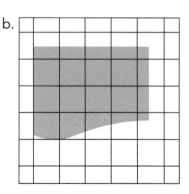

3. An interior decorator needs to purchase lace and wallpaper to remodel a rectangular table which has a width 3 m and length which is 3 times the width.

   a. How much lace should she buy to go around the edge of the table?

   b. How much wallpaper does she need to cover the table top?

4. Mr Joseph's square garden needs 28 m of fencing wire to fence it completely.

   a. What is the length of one side of the garden?

   b. What is the area of the garden?

   c. If he grows vegetables on a triangular patch which cuts his garden into two equal pieces, what is the area of his vegetable plot?

Now, try it out

1. **Find the answers to the following:**
   a. A square has one side 12 m.
      Area of square = _____
   b. A triangle has a base of 5 m and a height of 10 m.
      Area of triangle = _____
   c. A rectangle has an area of 32 sq metres and a length of 8 m.
      Width of rectangle = _____
   d. A triangle has a base of 3 cm and a height of 8 cm.
      Area of triangle = _____

2. **Consider the triangular plot below.**
   If shrubs cost $25 per square metre, what will it cost to fill the plot with shrubs?

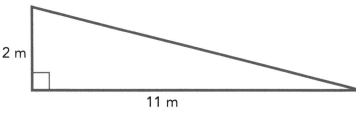

2 m

11 m

F Circumference & area of circles

Points to remember

- The distance around a circle is called the *circumference*.
- The point at the middle of a circle is called its *centre*.
- All points from the circumference of a circle are the same distance from the centre.
- A line which goes through a circle across its centre is called the *diameter*.
- A line which starts at the circumference of a circle and stops at its centre is called the *radius*.
- When finding area and circumference of circles the symbol π is equal to $\frac{22}{7}$ or 3.14.
- The formula for finding the circumference of a circle is $2\pi r$. This means $2 \times \pi \times radius$.
- The formula for finding the area of circle is πr^2. This means $\pi \times radius^2$ or $\pi \times radius \times radius$.

Example

A circular fruit cake has a radius of 15 cm. Using π as 3.14 find:

a. Its area

b. Its perimeter

Solution

Area = πr² = 3.14 x 15cm x 15cm = 3.14 × 225cm² = **706.5cm²**

Circumference = 2 x π x r or π x d = 2 × 15cm x 3.14 = 30cm x 3.14 = **94.2cm**

Exercise 6a

1. **Draw a circle and label** the following parts.

   • diameter AB   • centre C   • radius CD

2. **The wheel on a toy car has a radius of 5 cm. Using π as 3.14,**

   a. Find its circumference.   b. Find its area.

3. A caterer uses ribbon to create a design around a circular wedding cake. If the radius of the cake is 10 cm, how much ribbon is need for 3 complete turns around the cake? (Use π = 3.14)

Exercise 6b

1. **Complete the table. Remember the formula for the area of a circle is πr² and the formula for the circumference is 2πr. (Use π = 3.14).**

Radius	Diameter	Circumference (2πr)	Area (πr²)
2 cm			
5 cm			
10 cm			
12 cm			

2. **Draw the following circles and work out the circumference (use π = $\frac{22}{7}$).**

   a. a circle with radius of 2.5 cm

   b. a circle with radius of 6 cm

   c. a circle with diameter of 14 cm

3. In the diagram below circle X is inside of circle Y but has the same centre as circle Y. Using π = $\frac{22}{7}$, find:

a. the circumference of circle X

b. the circumference of circle Y

c. the area of circle X

d. the area of circle Y.

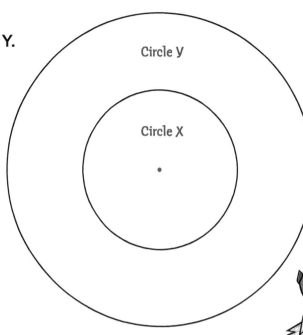

Circle Y

Circle X

Now, try it out

1. Cut out a circle from a manila sheet or some other hard paper material.

2. Use a piece of string to measure around the circle from one point to the other.

3. Measure the length of the string using your ruler and record your measurement.

4. Use your pencil to make a point at the centre of the circle. (Measure from the circumference of the circle to the centre on both sides to ensure that your point is at the centre.)

5. Measure the distance from the centre to the circumference of the circle.

6. Approximately how many times longer is the distance around the circle than the distance from the centre to the circumference?

7. Repeat for other circles of different sizes and record your results in the table below.

8. Does the proportion remain the same each time you measure?

Circle	Circumference in cm	Distance from the centre to the circumference in cm	Approximate number of times longer
1			
2			
3			
4			

(G) Capacity

Points to remember

- The amount of liquid a container can hold when it is full is called its *capacity*.
- A *litre* is the basic unit for measuring capacity.
- 1 litre = 1000 *millilitres* = 1000 cm³
- A container with a volume of 1000 cm³ has a capacity of 1 litre.

 Example 1

Change 5600 ml to litres.

Solution

5600 ml/1000 ml = 5.6 litres

 Example 2

If a box of juice has a volume of 3000 cm³, what is its capacity?

Solution

$\frac{3000}{1000}$ = 3 litres

Exercise 7a

1. Find the capacity of a cube with dimensions of 6 cm.

6 cm

2. The volume of a bottle of liquid multivitamins is 150 ml. Sandra takes a teaspoon full (which holds 5 ml of liquid) every day. In how many days will the bottle be empty?

3. Margaret bought a $3\frac{1}{2}$ litre bottle of passionfruit syrup. She used 1 litre of it in a fruit punch. She then gave each of her 4 classmates 250 ml to make juices for their school fundraising activity. How much of the passionfruit syrup remained? Give your answer in millilitres and litres.

Exercise 7b

1. In order to make fruit cocktail, Samantha needed 125 ml of fruit syrup, $\frac{3}{4}$ litres of fruit extract and 2.5 litres of soda water. How many litres of liquid does the recipe take altogether?

2. Martha's water bottle holds 2 litres of water. During her school's sports meet she drank 500 ml. She then gave her friend Sarah $\frac{1}{2}$ litre. How much water was remaining in the water bottle?

Now, try it out

Three containers of milk have the following capacities: 6 litres, 4 litres and 3.5 litres.

a. How many millilitres of milk would there be if all the milk was poured into one large container?

b. If the contents of a bottle with a capacity of 750 ml was added, what would be the total capacity?

c. If 2.5 litres of milk from the _original amount_ was used to make a cake, how many litres would be left?

H Telling and writing time

Points to remember

- Time can be measured using a 12-hour clock or a 24-hour clock.
- am and pm are used to show if the time is before noon or after noon.
- am refers to all times from after midnight to before noon and pm refers to all times after 12 noon up to midnight

 Example

What time is shown on the clock?

Solution

The time on the clock is 10:45 or 45 minutes after 10 or a quarter to 11.

Exercise 8a

1. **Use what you know about the units of time to work out the following.**

   a. 5 hours   = ----   seconds

   b. 2 hours   = ----   minutes   = ---- seconds

   c. 3 days   = ----   hours   = ---- minutes

   d. 7 h 24 min   = ----   min

2. **What part of 1 day is 2 hours?**

3. **How many seconds are in 1 hour?**

4. **Write the time as shown on the following clocks.**

a. 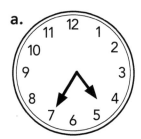   b. 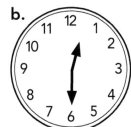   c.    d.

Exercise 8b

1. **The table shows the arrival and departure times of a school bus. Study the table and answer the questions.**

Day	Arrival	Departure
Monday	7:05 am	4:20 pm
Tuesday	7:20 am	4:00 pm
Wednesday	7:35 am	3:55 pm
Thursday	7:45 am	3:50 pm
Friday	7:50 am	3:50 pm

   a. On which day was the overall time at school the longest?

   b. On which day was the overall time at school the shortest?

   c. How much time elapsed between the arrival and departure times on Tuesday?

   d. How much time elapsed between the arrival and departure times on Wednesday?

   e. If school begins at 8:00 am, how early was the bus on Monday?

   f. If school ends at 3:45 pm, how late did the bus leave on Monday?

Now, try it out

How many minutes are there in one week? Show how you worked out your answer.

ⓘ 24-hour clock

Points to remember

- The hands of a clock always turn in a clockwise direction.
- Midnight to noon is 0:00–12:00 hours.
- 12 noon is 12:00 hours.

 Example

What are the following times using the 24-hour system?

a. 1 am   b. 1 pm   c. 2 pm   d. 7 pm

Solution

1 am is represented as 01:00

1 pm is represented as 13:00 hours (12 + 1)

2 pm is represented as 14:00 hours (12 + 2)

7 pm is represented as 19:00 hours (12 + 7)

Exercise 9a

1. **Write these times using the 24-hour system**

   a. 7:40 am   _____   b. 5:35 pm   _____

   c. 11:23 pm   _____   d. 12 midnight   _____

   e. 9:53 am   _____

2. **Complete the table.**

Time on 12-hour clock	Time on 24-hour clock
9:59 am	
	17:45
	19:05
7:12 am	
3:05 pm	

Exercise 9b

Give your answers using the 24-hour clock.

1. Susan left home for a party at 5:05 pm. She spent 15 minutes walking, then took a bus ride for $1\frac{1}{2}$ hrs. She then spent 20 minutes walking to her friend's house. What time did she arrive at the party?

2. A man began a task at 20:45. He finished it three hours later. What time was it when he finished?

3. Jerry started his walk to school at 6:35 am. He spent 1 hour, 15 minutes walking. What time did he arrive?

4. The time on a clock is 11:09 am. What is the time 8 hrs later?

Now, try it out

Write down ten different times of your own using a 12-hour clock and say whether each one is an 'am' or 'pm' time. Now write down the equivalent times using the 24-hour clock.

Points to remember
- 1 day = 24 hours
- 1 hour = 60 minutes
- 1 minute = 60 seconds

J Finding lengths of time

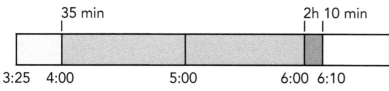 Example

You left school at 3:25 pm and arrived home at 6:10 pm.
How long did it take you to get home?

Solution

You need to find the difference in time between 3:25 pm and 6:10 pm.

Method 1:
3:25 + 35 = 4:00
4:00 + 2 h = 6:00
6:00 + 10 = 6:10 ⟶ add 35 min, 2 h
                        and 10 min = 2:45 h

Method 2:
6:10 − 3:25 ⟶ rewrite 6:10 as 5 h 70 min;
5:70 − 3:25   subtract minutes, then
                 subtract hours 2:45

Method 3:
Use a number line:

```
          35 min                          2h 10 min
          |                               |  |
  ┌───────┬───────────────┬──────────────┬──┬────────┐
  │       │               │              │▓▓│        │
  └───────┴───────────────┴──────────────┴──┴────────┘
 3:25   4:00            5:00           6:00  6:10
```

Method 4:
3:25 to 6:10 is almost 3 hours.
The difference between 25 and 10 is 15 so subtract 15 minutes from 3 hours.
The answer is 2:45 h or 2 h 45 min.

Exercise 10a

1. **Brian left home for school at 8:30 am. He ran for 10 min, walked another 15 min, then ran again for 10 min before he arrived. What time did he arrive at school?**

2. **How long is it in weeks and days from April 15 to July 10?**

3. **Jeffrey left for work at 7:30 am and arrived 15 minutes later. Clement left for work at 7:25 am and arrived 25 minutes later.**

   a. Who arrived at work first?

   b. If the official start time at work is 9:15 am, how long was Jeffrey at work before the official time?

4. The table shows a class timetable.

Subject	Start	Finish
Mathematics	8:30 am	10:00 am
English	10:00 am	12:15 pm
Lunch	12:15 pm	1:15 pm
Social Studies	1:25 pm	3:00 pm
Science	3:00 pm	4:15 pm

a. How much time is spent on mathematics?

b. Which subject is given the most time?

c. How many hours will a student be in the classroom on this day?

Exercise 10b

1. A ship was due to arrive at the Roseau Cruise Ship Berth at noon on Monday, but it arrived at 1:00 pm on Tuesday. How late was the ship?

2. If the time in the Caribbean is 5 hours behind that of London, what time is it in the Caribbean when it is 8:42 in London?

3. It takes a bus 5 hours 45 minutes to complete one leg of a journey. If it is allowed a $1\frac{1}{2}$ hour stop before beginning the return trip, how long does the entire journey take?

4. How many minutes are there between 12:00 noon and 9:38 pm?

5. Dionne spent 2 hours on her homework; she did the house chores in $1\frac{1}{2}$ hours and left home 1 hour later. If she left home at 2:30 pm, what time did she begin her homework?

Now, try it out
Game: *Time list*

Unscramble the letters in each cloud bubble to find different measurements of time.

ECSN OD

MTN HO

ECD AED

MTIE NU

YTU ERNC

(K) Calculating distance and average speed

Points to remember

- *Average speed* = distance ÷ time
- *Distance* = average speed × time
- *Time* = distance ÷ average speed
- Distance is measured in km, m and cm.
- Speed is distance covered in a given time.
- Time is measured in hours or minutes.
- Speed is measured in m/s, km/h or mph.

 ## Example 1

A car drives 300 miles in 2 hours. What is the average speed of the journey?

Solution

$$\text{Average speed} = \frac{\text{distance travelled}}{\text{time taken}} = \frac{300 \text{ miles}}{2 \text{ hrs}} = 150 \text{ mph}$$

 ## Example 2

A trip is made at an average speed of 20 km per hour. It took 4 hours to complete the journey. How long was the journey?

Solution

Distance = **average speed** x **time** = 20km x 4 = 80 km

Exercise 11a

1. If a 320 km trip is driven at an average speed of 40 km per hour, how long does the trip take?

2. **The graph shows the time taken for a trip. Use this information to answer the questions.**

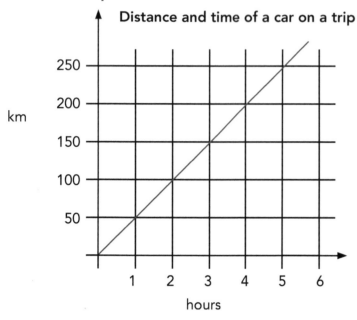

Distance and time of a car on a trip

km

hours

   a. How many km did the vehicle travel in 4 hours?

   b. About how long did it take to drive 250 km?

   c. What was the average speed of the vehicle?

3. **How much time would you save by driving 300 kilometres at 60 kilometres per hour (km/h) rather than at 50 km/h?**

4. **Find the distance travelled if you drive at 40 km/h for 30 minutes.**

Exercise 11b

1. **Complete the table.**

Average speed	Distance	Time
60 mph	180 miles	
24 mph		$3\frac{1}{2}$ hours
	90 miles	2 hrs
	120 miles	1 hr

2. **John travels on a bicycle with an average speed of 20 miles per hour. How far can he travel in 55 hours?**

3. An airplane travels 8000 miles in 4 hours. What is its average speed?

Now, try it out

Two cars are travelling the same distance at average speeds of 20 miles per hour and 80 miles per hour respectively. If the faster car reaches its destination in 2 hours, how long would the slower car take?

PRACTICE TEST 8 - MEASUREMENT

- Let's see how much you know

Section A

Select the letter of the correct answer to each question.

1. Which of these would be best measured in yards?

   [a.] a spoonful of salt [b.] a piece of meat [c] a ribbon tied around a gift box [d.] a bottle of oil   Ⓐ Ⓑ Ⓒ Ⓓ

2. The length of the line above is approximately

   [a.] 5 mm [b.] 12 mm [c.] 5 cm [d.] 12 cm   Ⓐ Ⓑ Ⓒ Ⓓ

3. Which shape has the shortest distance around it?

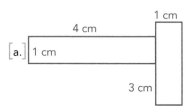

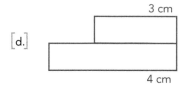

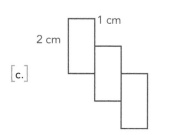

Ⓐ Ⓑ Ⓒ Ⓓ

4. Which is the heaviest?

   [a.] 1 kg [b.] 1 milligram [c.] 1 centigram [d.] 1 g   Ⓐ Ⓑ Ⓒ Ⓓ

5. A square room has an area of 100 m². What is its perimeter?

   [a.] 100 m [b.] 40 m [c.] 20 m [d.] 400 m   Ⓐ Ⓑ Ⓒ Ⓓ

6. A storage tank holds 156 gallons of coconut oil. The oil is imported in quart containers. It takes 4 quarts to fill a gallon container. How many quarts will fill the tank?

   [a.] 624 quarts [b.] 152 quarts [c.] 160 quarts [d.] 312 quarts   Ⓐ Ⓑ Ⓒ Ⓓ

7. A square box has a perimeter of 72 cm. What is its area?

   [a.] 72 cm² [b.] 288 cm² [c.] 154 cm² [d.] 324cm²   Ⓐ Ⓑ Ⓒ Ⓓ

8. What is the perimeter of an octagon with sides 11 metres long?

   [a.] 99 metres [b.] 66 metres [c.] 77 metres [d.] 88 metres   Ⓐ Ⓑ Ⓒ Ⓓ

9. The total mass of six identical crates of drink is 42 kg. What is the mass of each crate?

   [a.] 7 kg [b.] 24 kg [c.] 252 kg [d.] 42 kg   Ⓐ Ⓑ Ⓒ Ⓓ

10. Which of these is the approximate weight of a car?

    [a.] 1 centigram [b.] 1 milligram [c.] 1 kg [d.] 1 tonne   Ⓐ Ⓑ Ⓒ Ⓓ

11. Ten poles are 3 metres apart. What is the distance from the first to the last?

    [a.] 30 metres   [b.] 27 metres   [c.] 3 metres   [d.] 90 metres   Ⓐ Ⓑ Ⓒ Ⓓ

12. Sammy's juice bottle holds 3 litres of juice. He drank 750 ml at school and 500 ml when he arrived home. How much juice was remaining in the bottle?

    [a.] 1 litre 750 ml   [b.] 2 litres   [c.] 500 ml   [d.] 3 litres 750 ml   Ⓐ Ⓑ Ⓒ Ⓓ

13. An aircraft was due to arrive at the airport at 3:00 pm on Saturday, but arrived at 5:35 pm on the same day. How late was the aircraft?

    [a.] 35 minutes late   [b.] one hour 25 minutes late   [c.] 1 hour 35 minutes late   [d.] 2 hours 35 minutes late   Ⓐ Ⓑ Ⓒ Ⓓ

14. How many minutes are there between 2:05 pm and 3:30 pm on the same day?

    [a.] 85 minutes   [b.] 60 minutes   [c.] 80 minutes   [d.] 25 minutes   Ⓐ Ⓑ Ⓒ Ⓓ

15. Dion went for a jog. She jogged 72 km in 12 minutes. How far will she jog in 72 minutes at the same speed?

    [a.] 432 km   [b.] 420 km   [c.] 84 km   [d.] 12 km   Ⓐ Ⓑ Ⓒ Ⓓ

Section B

Answer the following.

1. A fish tank has dimensions of 12 cm x 10 cm x 9 cm. What is its volume?

    _____

Study the table below and answer the questions.

Student	Date of birth
Zach	June 1, 1996
Jerome	July 1, 1995
Micah	August 10, 1998
Tom	May 12, 1993
Cindy	August 13, 1994

2. Which is the youngest student?

    _____

3. Which is the oldest student?

    _____

4. Sandra is decorating her room for a birthday party. She purchased 1 m of ribbon. How many 5 cm cuts can she get in order to make her design?

    _____

5. The daytime temperature on a farm is 35°C. At nights the temperature drops to 9 degrees. What is the difference between temperature during the day and the night-time temperature?

    _____

GEOMETRY AND SPATIAL SENSE

The use of Geometry began in Ancient Egypt before 1700 BC. However, around 300 BC the Greeks formally established the concept. Geometry explores spatial relationships and concepts.

In this chapter you will review and reinforce your understanding of:

- shapes and their properties
- symmetry
- congruence
- transformation.

What you need to know

* An angle is formed when two straight lines meet at a point.

* Angles on the inside of a shape are called *interior angles*.

* Angles on the outside of a shape are called *exterior angles*.

(A) Naming and classifying angles and lines

Points to remember

 Angles are measured in degrees using a *protractor*. This represents the amount of turn.

 Right angle, $\frac{1}{4}$ turn = 90°

 Straight angle, $\frac{1}{2}$ turn = 180°

 A *complete turn*, 360°

 An acute angle is less than 90°.

An obtuse angle is more than 90° but less than 180°.

 A reflex angle is greater than 180° but less than 360°.

- A straight line drawn from North to South or South to North is a vertical line.
- A straight line drawn from East to West or from West to East is a horizontal line.
- Perpendicular lines are those which intersect each other at right angles
- Parallel lines are those which are the same distance apart for their entire length. Parallel lines never meet.

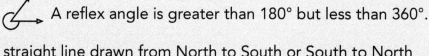

Exercise 1a

1. a. Draw a horizontal line.   b. Draw a vertical line.
    c. Draw two lines which are parallel.

2. a. Name the angles you see in each figure.
    b. State whether they are acute, obtuse, reflex or right angles.

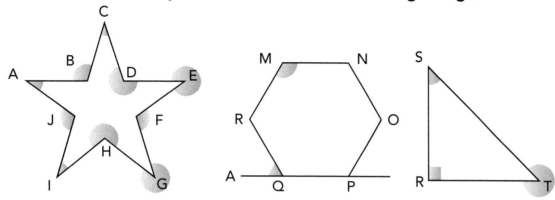

3. Identify the type of angle which is described in each of the cases below.

    a. At 3'o clock the hands of the clock form a _____ angle.

    b. A _____ angle is greater than 90° but less than 180°.

    c. If the minute hand of a clock moves from 15 minutes past the hour to 25 minutes past the hour a _____ angle is formed.

Exercise 1b

1. **How many interior angles can be found in this shape?**

2. **Draw each of the following angles.**
    a. an acute angle
    b. a right angle
    c. an obtuse angle
    d. a reflex angle

3. **Use a ruler and a protractor to draw the following angles.**
    a. an angle measuring 45°
    b. an angle measuring 20°
    c. an angle measuring 55°
    d. an angle measuring 80°

Now, try it out

Look around your classroom. Identify ten objects which form angles. Now group these angles into three categories namely those forming right angles, those greater than right angles and those less than right angles. Represent the information in the form of a table.

(B) Properties of 2-dimensional shapes

Points to remember

- A *polygon* is a flat (or closed) shape made from three or more line segments.

Examples of polygons are:

- Squares, rectangles, triangles, trapezium, etc.

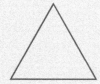 

Exercise 2a

1. **Complete this table.**

Shape	Number of sides	Number of angles
triangle		
square		
rectangle		
pentagon		
hexagon		

2. a. **What is the difference between a square and rectangle?**

   b. **Lines that intersect (meet) each other at right angles are called *perpendicular lines*. Which of the polygons above have perpendicular sides?**

   c. **Which of the polygons above have parallel sides?**

3. **A 4-sided polygon is called a *quadrilateral*. Give three examples of quadrilaterals.**

Exercise 2b

1. **Name the following types of triangle.**

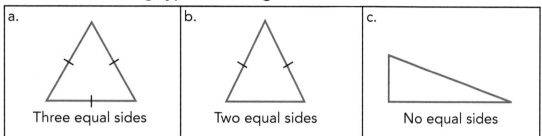

a.	b.	c.
Three equal sides	Two equal sides	No equal sides

2. **Look carefully at the diagram below and complete the table.**

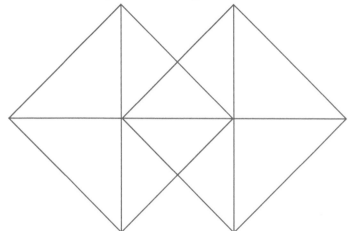

Number of triangles	Number of rectangles	Number of squares

Now, try it out

Use the following 2D shapes to create patterns or objects of your own. You may use a shape more than once. You may also colour the shapes if you wish.

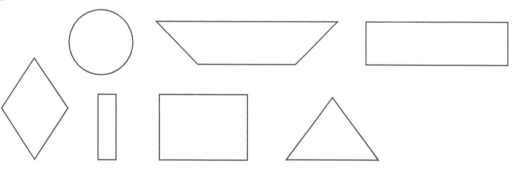

ⓒ 3-dimensional shapes

Points to remember

- A *solid* is an object that takes up space.
- Solids are called *3-dimensional* (3D) shapes.
- This means that solids have length, width and height.

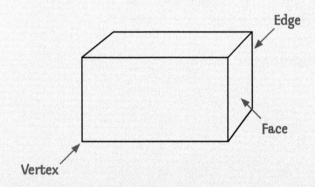

Edge

Face

Vertex

A *face* is a plane surface.
An *edge* is the line where two faces meet.
A *vertex* is the corner where edges meet.
The plural of vertex is *vertices*.

Exercise 3a

1. **Complete the table. The first one is completed as an example.**

Shape	Name	Number of faces	Number of vertices	Number of edges
	cuboid	6	8	12
	cube			
	cylinder			
	cone			
	sphere			
	triangular pyramid			

2. **Look around the classroom and try naming some objects that look like solids.**
   Discuss your selections with a friend.

Exercise 3b

1. **Write the name of the shape which each of these objects represents:**

   a. a matchbox b. a lime

   c. a can of evaporated milk d. a dice

2. **Give examples of two items in the environment which form each of these shapes:**

   a. a sphere b. a cone

   c. a cube d. a cuboid

Now, try it out

Cut through the centre of the following solids and name the shape that you see when looking at the centre.

a. an orange
b. an ice cream cone
c. a small matchbox

D Nets of solids

 Points to remember

- A *net* is a flat pattern that folds into a solid.
- Opening the sides of a solid gives the flat pattern.

Example

What shape would you get if you folded this net into a solid?

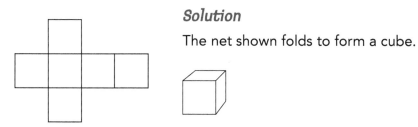

Solution

The net shown folds to form a cube.

Exercise 4a

1. **Which pattern will work as a net for an open cube? Which will not? Explain your answer.**

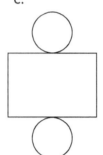

2. Name the solid you can make from each of these nets.

a.   b.   c.

 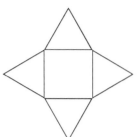

Trace each net on hard paper and fold along the solid lines. Tape as needed to form the solid.

Exercise 4b

1. Using squared paper, draw diagrams to represent the net of a cube.

2. Using plain paper, draw diagrams to show nets of the following solids.

    a. a cone

    b. a cylinder

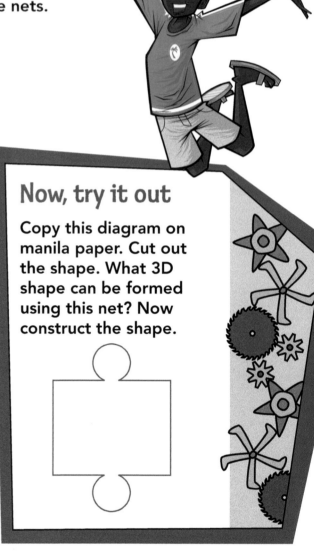

Now, try it out

Copy this diagram on manila paper. Cut out the shape. What 3D shape can be formed using this net? Now construct the shape.

E Similarity and congruence

Points to remember

- Figures that are the same size and shape are *congruent*.
- Sometimes you can have figures that are the same shape but different sizes. These are called *similar figures*.

Example

Congruent shapes are an exact copy of each other and have the same features. The following shapes are congruent;

Exercise 5a

1. **Look at these figures and identify the two pairs of congruent figures.**

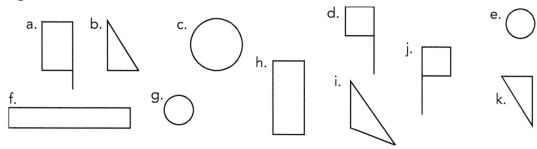

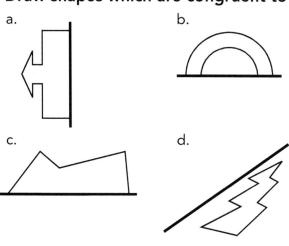

Exercise 5b

1. **Draw shapes which are congruent to these.**

a.

b.

c.

d.

2. **Draw a solid which has 1 vertex, 1 acute angle and a circular face. What object does your shape look like?**

Now, try it out

Work with a partner. Draw a shape and allow your partner to draw a shape which is congruent to the one which you drew. Reverse the activity.

(F) Drawing symmetrical shapes

Points to remember

■ A *line of symmetry* divides an object into two identical parts.

 Example

Does this shape have symmetry?

Solution

This shape does have symmetry.

If you draw a vertical line through the centre, the pattern on each side will match.

The line is called a *line of symmetry*.

Exercise 6a

1. **Predict how many lines of symmetry each shape has. Trace the shape, cut it out and fold it. Were you correct?**

a.　　　　　　　　　b.　　　　　　　　　c.

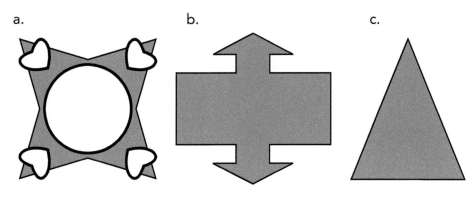

2. State whether the line drawn in each figure is a line of symmetry for that figure.

a.    b. 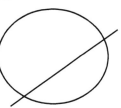   c.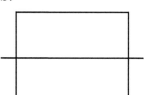

Exercise 6b

1. Look at the word 'RHONDA' and write the letters in a larger size.

   a. Which letters have a line of symmetry?

   b. If a letter has more than one line of symmetry, draw the lines.

Now, try it out

1. Using matchsticks, toothpicks or coloured short straws make the following polygons ('poly' means many, 'gon' means sides):
   a. a 9-sided figure with 2 right angles
   b. a 4-sided figure with 1 pair of parallel sides
   c. a 5-sided figure with 2 right angles

2. Draw a regular polygon for each shape and then identify the lines symmetry for each. Complete the table and discover the pattern.

Number of sides	3	4	5	6	7	8	n
Number of lines of symmetry							

PRACTICE TEST 9 - GEOMETRY AND SPATIAL SENSE

- Let's see how much you know

Section A

Select the letter of the correct answer to each question.

1. **Which of these pairs of lines forms an angle?**

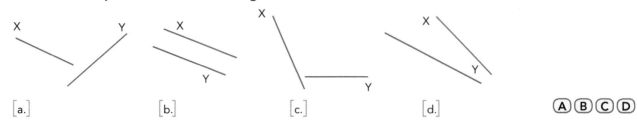

   [a.]   [b.]   [c.]   [d.]   Ⓐ Ⓑ Ⓒ Ⓓ

2. **In the figure below,**

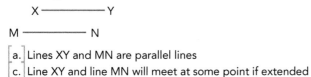

   [a.] Lines XY and MN are parallel lines   [b.] Line XY intersects line MN
   [c.] Line XY and line MN will meet at some point if extended   [d.] Line XY and line MN form an angle   Ⓐ Ⓑ Ⓒ Ⓓ

3. **Identify the vertical line.**

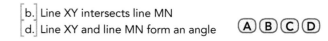

   [a.]   [b.]   [c.]   [d.]   Ⓐ Ⓑ Ⓒ Ⓓ

4. **Which of these shows a reflex angle?**

   [a.]   [b.]   [c.]   [d.]   Ⓐ Ⓑ Ⓒ Ⓓ

5. **Which of these shows an obtuse angle?**

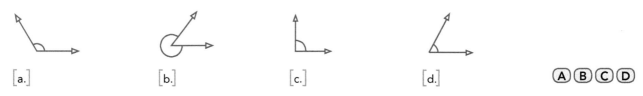

   [a.]   [b.]   [c.]   [d.]   Ⓐ Ⓑ Ⓒ Ⓓ

6. **Which of these shapes are congruent?**

   [a.]   [b.]   [c.]   [d.]   Ⓐ Ⓑ Ⓒ Ⓓ

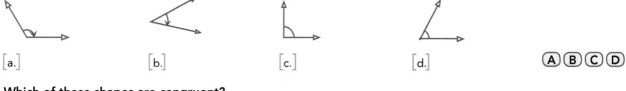

7. **How many sides can be found in an octagon?**
   [a.] 5   [b.] 6   [c.] 7   [d.] 8   Ⓐ Ⓑ Ⓒ Ⓓ

8. **In the shape below the angle marked A is a(n)_____.**
   [a.] Right angle   [b.] obtuse angle   [c.] acute angle   [d.] reflex angle   Ⓐ Ⓑ Ⓒ Ⓓ

9. Which of these is a right-angled triangle?

[a.]   [b.]   [c.]   [d.]   Ⓐ Ⓑ Ⓒ Ⓓ

10. Susan drew a shape on a piece of paper. The shape had 2 acute angles, 2 obtuse angles, and 4 equal sides. Which shape could she have drawn?

[a.] pentagon   [b.] square   [c.] equilateral triangle   [d.] rhombus   Ⓐ Ⓑ Ⓒ Ⓓ

11. In the shape below, angle 'A' is approximately

[a.] 20°   [b.] 60°   [c.] 120°   [d.] 180°   Ⓐ Ⓑ Ⓒ Ⓓ

12. Congruent shapes have
[a.] only corresponding sides equal
[b.] only corresponding angles equal
[c.] corresponding sides equal but opposite sides different
[d.] corresponding sides and angles equal   Ⓐ Ⓑ Ⓒ Ⓓ

13. A basketball is an example of what shape? _____
[a.] globe   [b.] sphere   [c.] cone   [d.] hexagon   Ⓐ Ⓑ Ⓒ Ⓓ

14. Which of these represents the net of a cylinder?
[a.]   [b.]   [c.]   [d.]   Ⓐ Ⓑ Ⓒ Ⓓ

15. How many lines of symmetry can be found in this shape?

[a.] one   [b.] two   [c.] three   [d.] four   Ⓐ Ⓑ Ⓒ Ⓓ

Section B

1. Draw the following shapes:
[a.] pentagon   [b.] hexagon   [c.] octagon

2. Draw the following:
[a.] An obtuse angle   [b.] A reflex angle

3. What angle is formed when the hands of a clock strike 6.30pm?
_____

4. What angle is formed when the hands of a clock strike 3:00pm?
_____

5. Draw two shapes which are congruent.

10 DATA HANDLING AND PROBABILITY

Data collection and handling have been used since biblical times. During this period, shepherds would use statistical methods to account for their flock. Today, we collect and use data in almost every thing which we do.

In this chapter you will review and extend your knowledge of:

- data collection and interpretation
- calculating averages from a group of data
- the performance of probability experiments
- reading and interpreting data from various graphs, charts and tables.

What you need to know

1. Some of the most useful ways of collecting data are by *observation*, *interviews* and *questionnaires*.

2. Information can be organized by using various statistical diagrams. Four of the most frequently used types are *bar graphs*, *pie charts* (or *circle graphs*), *line graphs* and *pictographs*.

(A) Organizing data using tables and bar graphs

Points to remember

- People collect data for different reasons and from different sources. Information can be organized by using tally charts, frequency tables and bar graphs. For example:

  1. an election poll to determine which political party people prefer

  2. a survey to find out which car is the best buy

  3. weather forecast predictions

  4. students' performance on class test.

 Example

White, blue, yellow, pink, green, black, red, yellow, pink, white, white, black, red, yellow, red, green, white, green, blue, brown, pink, white, pink, green, pink, red, pink, yellow, blue, white, black, green, red, pink, red, brown, pink, white

The information above shows the favourite colours of students in **Grade 6B**. Draw
a. A frequency table
b. A bar graph to represent the information.

Solution

Frequency Table

Colour	Tally	Total
White	ⅲℍ II	7
Blue	III	3
Green	IIII	5
Black	III	3
Red	ℍ I	6
Yellow	IIII	4
Brown	II	2
Pink	ℍ III	8

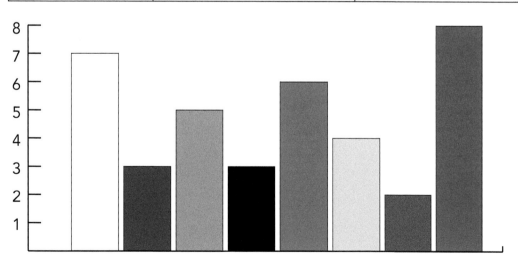

Exercise 1a

1. **The information below shows the months in which students from a Grade 6 class were born.**

   January: Jane, John, Mark, Jerry

   February: Mark, Jeff, Esther, Richie, Sandra

   March: Tony, Robert, Bob, Catherine

   April: Nadia, Pam

   May: Mary, Nick, Alvin, Brian, Paula, Dave

   June: Jessica, Joanne, Maria, Susan, Lisa

   July: Garvin, Caril, Tim, Nelly, Eddie, Kurt, Ralph, Mattie

   a. Complete the following frequency table using the information presented.

   b. Draw a bar graph to represent this information.

Months	Frequency	Total
January	IIII	4
February		
March		
April		
May		
June		
July		

2. **Your class teacher would like to find out the most popular subjects in her class.**

   a. How could she collect this information?

   b. In what ways could she represent this information?

   c. Conduct a survey to collect this information and present it by using a form of representation.

Exercise 1b

1. **Count the number of times each vowel appears in the verse below.**

   a. Record your results in the table.

   b. Draw a bar graph using the data collected.

   Real, real, Maths is real to me
   I know it and
   I can't live without it
   That is why I love it so
   Maths is real to me!

Vowel	Tallies	Total
A		
E		
I		
O		
U		

2. **Display the following data in a frequency table: 5, 1, 4, 6, 2, 6, 4, 5, 1, 3, 2, 6, 4, 5, 4, 6**

Number						
Frequency						

3. **The chart shows countries that students would like to visit.**

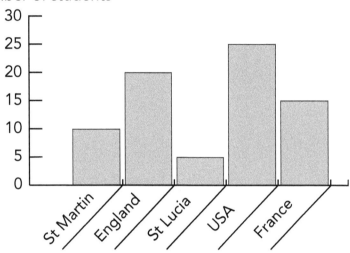

Number of students

a. Which country is *most* popular?

b. Which country is *least* popular?

c. How many students in total would like to visit St Martin and England?

d. How many more students wanted to go to the USA than to St Lucia?

e. How many students were in the class altogether?

Now, try it out

Work in groups of three students. Each group should select one class within the school to undertake a small survey. Find out one of these:

a. the months of the year in which students were born

b. the shoe sizes of students

c. students preferred colour

Use the information collected to construct a frequency table and then a bar graph.

B Organizing data using pie charts, pictographs and line graphs

Points to remember

1. Pie charts are used to show the parts of a whole.
2. Often the parts in a pie chart are expressed as percentages.
3. The sum of all the percentages in a pie chart is 100%.
4. Pictographs are used to show information using pictures or symbols.
5. Line graphs are used to show trends that change over time, for example temperature or rainfall.

Example

Tracey received $100.00 as a birthday gift from her parents.
The table below shows how she spent the cash received.
Draw a pie chart to represent the information.

Item	Amount Spent
Perfume	$15.00
Snacks	$5.00
Books	$24.00
Clothing items	$30.00
Savings	$26.00

Solution

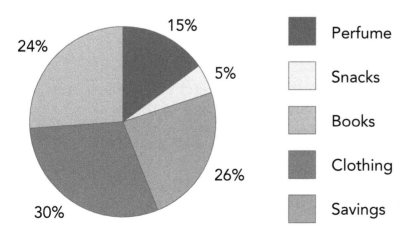

Exercise 2a

1. **The pie chart shows how Mr James spent his monthly salary of $3000.**

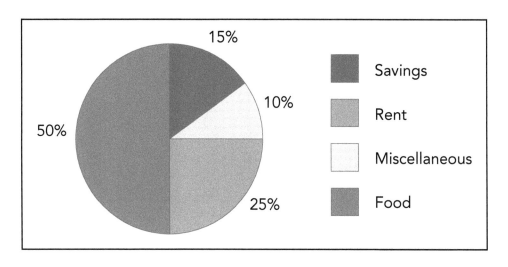

   How much does Mr James spend on

   a. rent   b. food   c. miscellaneous items?

2. **The pictograph shows the number of floor tiles laid in a kitchen from Saturday to Tuesday.**

Saturday	☺☺☺☺☺☺☺
Sunday	☺☺☺☺
Monday	☺☺☺
Tuesday	☺☺☺☺☺☺☺☺

☺ represents 20 floor tiles

   a. How many tiles were laid on each day?

   b. How many tiles were laid altogether?

   c. How many more tiles were laid on Saturday than on Monday?

3. **The line graph shows the average rainfall during different months in a given year. Use this information to find the difference between the average rainfall in March and in November.**

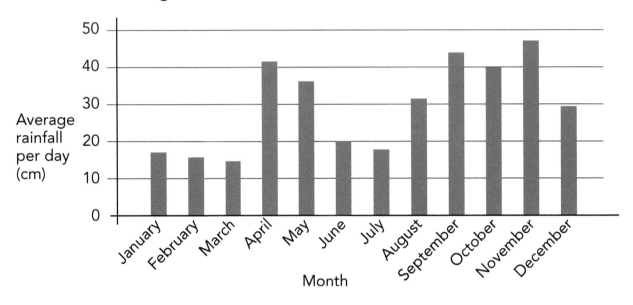

4. **The table shows tourist arrivals in some Caribbean islands for the month of December 2010. Study the table and answer the questions.**

Country		Tourist arrivals
St Lucia	(St. Lu)	30 000
Dominica	(Dom)	15 000
St Vincent and the Grenadines	(St. Vin)	11 000
Montserrat	(Mont)	2 000
Anguilla	(Ang)	16 000
Antigua and Barbuda	(Ant)	35 000
Grenada	(Gren)	18 000
St Kitts and Nevis	(St. Kits)	9 000

a. Which country had the highest number of tourist arrivals?

b. Which country had the fewest arrivals?

c. How many more tourists arrived in Dominica than Montserrat?

d. Draw a bar graph to represent the data using the coutry codes provided.

Exercise 2b

1. **The pie chart shows the food choices of students in a class. Study the diagram and answer the questions.**

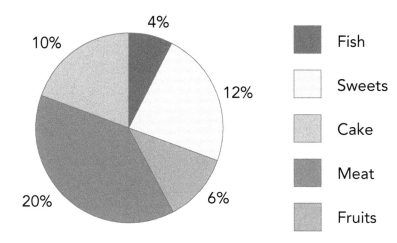

4%
10%
12%
20%
6%

Fish
Sweets
Cake
Meat
Fruits

a. Which food is the most popular?

b. Which is best preferred between sweets and fruits?

c. Which food is least preferred between cake and fish?

d. How many students were in the class altogether?

e. Complete the diagram below using *one smiling face* to represent *two* students.

Food choices of students in a class	Number of students
Fish	
Sweets	
Cake	
Meat	
Fruits	
KEY ☺ represents 2 students	

2. **The list below gives the makes of vehicles driving past a school during a 30-minute period.**

Cars: Toyota, Nissan, Toyota, Mitsubishi, Toyota, Nissan, Mazda, Nissan, Toyota

Buses: Nissan, Mitsubishi, Nissan, Mitsubishi, Mitsubishi, Toyota, Nissan, Nissan

Vans (4WD): Honda, Suzuki, Honda, Mitsubishi, Suzuki, Honda, Toyota, Honda

a. Use the information to complete the tables below.

(i)

Vehicle make	Frequency
Toyota	
Mitsubishi	
Honda	
Nissan	
Mazda	
Suzuki	

(ii)

Vehicle type	Frequency
Car	
Van (4WD)	
Bus	

b. Which type of bus was seen most often?

c. Which type of 4WD was seen most often?

d. Which type of bus was seen least often?

e. Which type of 4WD was seen least often?

3. **A basketball tournament was held at the St Peter's community centre. The line graph shows the attendance for the first seven months. Look at the graph and write at least two sentences based on the information it is showing.**

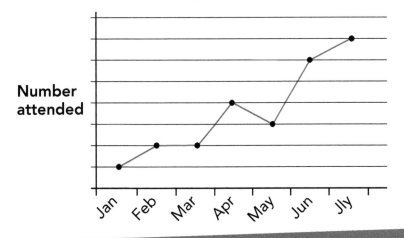

Number attended

Jan Feb Mar Apr May Jun Jly

Now, try it out

Use the information below to construct a pie chart. You may use different colour codes as a key to represent your information.

At a school's sports meet students wore the following house colours.

- 25 children wore yellow T-shirts.
- 15 children wore blue T-shirts.
- 22 children wore green T-shirts.
- 28 children wore red T-shirts.
- 24 children wore white T-shirts.
- 19 children wore other colours combined.

© Averages: mean, median, mode

Points to remember

1. We use three types of measure in simple statistics to describe the average of a list of numbers.
2. These measures of central tendency are called the *mean*, *median* and *mode*.
3. The *range* measures how scattered the data is.

 Example

Measures	Example
Mean The mean equals the sum of values divided by the number of values. $\text{mean} = \frac{\text{sum of values}}{\text{number of values}}$	**Example** Numbers: 80, 70, 85, 90, 70 $\text{Mean} = \frac{80+70+85+90+70}{5}$ $= \frac{395}{5} = 79$
Median To find the median, write the numbers in order from smallest to largest. The median is the middle number. If there are an even number of data values then there will be two 'middle' numbers. Take the mean of these two values.	**Example 1** Numbers: 70, 70, 80, 85, 90 Median = 80 **Example 2** Numbers: 70, 70, 80, 82, 85, 90 $\text{Median} = \frac{80+82}{2} = 81$
Mode The mode is the number that occurs the most times. A set may have more than one mode.	**Example 1** Numbers: 70, 70, 80, 85, 90 Mode = 70 **Example 2** Numbers: 70, 70, 80, 85, 85, 90 There are two modes: 70 and 85
Range The range gives an indication of how scattered the data is. It is the difference between the least and greatest values.	**Example** Numbers: 70, 70, 80, 85, 90 Range = 90 – 70 = 20

Exercise 3a

1. **Find the mean, median, mode and range for each set of numbers.**

   a. 30, 45, 10, 70, 43, 24, 30

   b. 75, 62, 54, 68, 75, 62, 72, 75, 69

   c. 3, 4, 1, 1

   d. 9, 3, 6, 4, 3, 10, 7

2. **The following are the scores of a group of students in a Math quiz.**

   85, 86, 45, 68, 70, 90, 35, 55, 50, 45, 87, 90, 48, 58, 80, 75, 74, 82, 84, 60, 65, 66, 53

   a. Find the mode, median and mean scores.

   b. What is the range of the scores?

Exercise 3b

1. **In a high jump competition, athletes cleared the following heights in feet.**

Franklyn:	9.5	8.6	9.5	8	7	8.5	9
Brian:	8	8	7.5	6	7	9	9.5
Ricky:	6	6	7	7.5	7	8	8.5
Amos:	7	7	7	7	8	8.5	9

   a. What was the mode of the scores for each of the athletes?

   b. What was the median score for each athlete?

   c. What was the mean score for each athlete?

   d. Work out the range of the scores for each athlete.

   e. How do you think the judges should identify the overall winner in the high jump competition?

Now, try it out

Work with a partner. Find out the ages (in years) of all the students in your class. Now find the mode, median and mean age of the students.

(D) Probability

Points to remember

1. *Probability* predicts the chance of something taking place.
2. Probability may be written as a fraction, a decimal or a percentage and ranges in value from 0 to 1.

   Probability = $\frac{\text{number of successful outcomes}}{\text{total number of outcomes}}$

 Example

What is the probability of tossing a coin and getting a head?

Solution

Divide the number of heads by the total number of outcomes:

$\frac{\text{Number of heads}}{\text{Total number of outcomes}} = \frac{1}{2}$

The probability is $\frac{1}{2}$ or 0.5 or 50%.

Exercise 4a

1. **There are 24 boys and 26 girls in class. One student is to be chosen for a prize gift.**

   a. What is the probability that a boy will be selected?

   b. What is the probability that a girl will be selected?

2. **A boy has a bag of 10 marbles: 3 blue, 5 green and 2 red.**

   He pulls a marble from the bag without looking. What is the probability that he will pull a red marble from the bag?

3. **What is the probability of choosing the letter 'U' from the word 'SQUARE', if all the letters are on bits of paper, turned face down and mixed up?**

Exercise 4b

1. **A die is rolled. What is the probability of getting:**

   a. a number less than 5   b. an even number?

2. **The following represents the inventory of the New Fashion shoe store for the year 2010.**

Shoe size	5	6	7	8	9	10	11	12
Number of pairs of shoes sold	30	35	25	40	55	50	65	20

a. Find the mean shoe sales for the year.

b. Which shoe size was closest to the mean sales?

c. Which shoe size was furthest from the mean sales?

d. Write at least two questions that can be answered by using the data in the table.

Now, try it out
Game: *Hits and misses*

Play with a partner.

1. Make a large copy of the circle. Colour each section.

2. Place the copy in a corner of your classroom.

3. Stand just over 1 metre from the circle.

4. Use a coin or button to toss onto the circle.

5. Each person has a total of 10 tosses.

6. Discard tosses that do not land on the circle.

7. Keep a tally of your results.

After playing the game, answer the following questions:

a. Which colour did you hit the most and which colour did you hit the least?

b. Is it more or less likely that you will hit one colour than another?

PRACTICE TEST 10 - DATA HANDLING AND PROBABILITY

- Let's see how much you know

Section A

Select the letter of the correct answer to each question.

The following table represents the favourite sports of students at a school.

Sport	Number of students
Football	60
Netball	45
Volley-ball	40
Basketball	80
Cricket	75
Boxing	10
Tennis	25

1. **Which is the most preferred sport?**

   a. football   b. basketball   c. netball   d. cricket   Ⓐ Ⓑ Ⓒ Ⓓ

2. **Which is the least preferred sport?**

   a. netball   b. volley-ball   c. tennis   d. boxing   Ⓐ Ⓑ Ⓒ Ⓓ

3. **What is the total number of students who prefer netball and volley-ball?**

   a. 105   b. 100   c. 95   d. 85   Ⓐ Ⓑ Ⓒ Ⓓ

4. **How many more students prefer cricket than tennis?**

   a. 50   b. 45   c. 55   d. 65   Ⓐ Ⓑ Ⓒ Ⓓ

5. **How many more students prefer cricket than football?**

   a. 75   b. 60   c. 40   d. 15   Ⓐ Ⓑ Ⓒ Ⓓ

6. **What is the difference between the number of students who prefer the most popular sport and the number of students who prefer the least popular sport?**

   a. 80   b. 70   c. 60   d. 10   Ⓐ Ⓑ Ⓒ Ⓓ

7. **What was the total number of students surveyed?**

   a. 450   b. 400   c. 335   d. 300   Ⓐ Ⓑ Ⓒ Ⓓ

The following represents scores obtained by students in a Science quiz.

85	40	45	50	75	90	35	40	50	55	60
40	35	70	75	70	40	50	45	90	95	65

8. **What is the mode of the scores?**

   a. 35   b. 40   c. 50   d. 90   Ⓐ Ⓑ Ⓒ Ⓓ

9. **What is the median of the scores?**

   a. 40   b. 50   c. 52.5   d. 55   Ⓐ Ⓑ Ⓒ Ⓓ

10. **What is the mean of the scores?**

   a. 59   b. 70   c. 80   d. 90   Ⓐ Ⓑ Ⓒ Ⓓ

11. **How many students received more than ten points higher than the mean?**

   a. 5   b. 8   c. 10   d. 12   Ⓐ Ⓑ Ⓒ Ⓓ

12. **How many students received more than ten points lower than the mean?**

   a. 5   b. 8   c. 10   d. 14   Ⓐ Ⓑ Ⓒ Ⓓ

13. How many students took part in the quiz?

[a.] 22   [b.] 23   [c.] 24   [d.] 25   Ⓐ Ⓑ Ⓒ Ⓓ

Study the table and answer the questions.

Make of vehicle	Sales in 2010
Toyota	54
Mitsubishi	50
Honda	34
Nissan	65
Suzuki	25

14. What was the total number of vehicles sold for 2010?

[a.] 65   [b.] 54   [c.] 200 per month   [d.] 228   Ⓐ Ⓑ Ⓒ Ⓓ

15. What was the average number of vehicles sold in 2010?

[a.] 12   [b.] 20   [c.] 45.6   [d.] 25   Ⓐ Ⓑ Ⓒ Ⓓ

Section 2

Answer the following.

The chart shows how Pam spent her month's earnings.

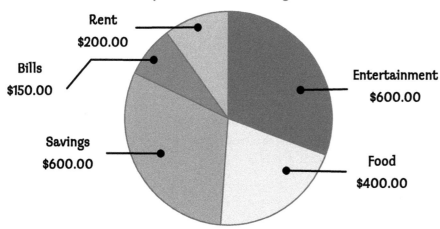

1. How much more was spent on entertainment than food?

_____

2. What was the difference between Pam's expenditure on entertainment and her expenditure on bills?

_____

3. How much did Pam spend on food and bills together?

_____

4. On which two items did Pam spend a total of $350.00?

_____

5. How much money did Pam receive for the month?

_____

CHAPTER 1 NUMBER IDEAS AND RELATIONS

Exercise 1a

1. a. 70   b. 80 000
2. a. one hundred and eighty thousand, nine hundred and twenty dollars
   b. April
3. a. three thousands   b. in the hundreds position
4. a. 60 000   b. 100 000   c. 40

Exercise 1b

1.

2. 100
3. 10
4. 1 734 519

 Now, try it out   (1A)

a. 1000 days   b. 2000 days
c. 5000 days   d. 10 000 days

Exercise 2a

1. a. 90 000 + 9000 + 200 + 30 + 4
   b. 700 000 + 40 000 + 5000 + 300 + 90 +1
   c. 1 000 000 + 800 000 + 60 000 + 3000 + 400 + 2
   d. 1 000 000 + 200 000 + 40 000 + 3000 + 400 + 20 + 1
2. a. 10^3 b. 10^4 c. 12×10^3 d. $14 000 = 14 \times 10^4$

Exercise 2b

1.

Expanded form	Power of 10	Standard notation
(7 × 10 000) + (5 × 1000) + (4 × 100) + (3 × 1)	$(7 \times 10^4) + (5 \times 10^3) + (4 \times 10^2) + 3$	75 403
1 000 000 + 800 000 + 600 + 70 + 4	$10^6 \times (8 \times 10^5) + (6 \times 10^2) + 7 \times 10 + 4$	1 800 674
20 000 + 3000 + 500 + 70 + 8	$(2 \times 10^4) + (3 \times 10^3) + (5 \times 10^2) + 7 \times 10 + 8$	23 578
(1 × 10 000) + (4 × 1000) + (5 × 100) + (3 × 10) + (3 × 1)	$10^4 + 4 \times 10^3 + 5 \times 10^2 + 3 \times 10 + 3$	14 533

2. a. 25 000   b. 2 000   c. 150 000   d. 12 000

 Now, try it out   (1B)

Students to identify their own numbers

Exercise 3a

1. a. 31 490   b. 31 500   c. 31 000
2. $79 000
3. 457 500

Exercise 3b

1. a. St Lucia   b. Dominica   c. Grenada
   d. St Kitts/Nevis and Grenada   e. 255 200
2. a. 456 700   b. 178 200   c. 34 400

 Now, try it out   (1C)

Rounded to nearest 10 000 = 20 000

Rounded to nearest 1000 = 20 000

Rounded to nearest 100 = 19 900

Exercise 4a

1. a. 51   b. 59   c. 19   d. 54   e. 44
2. a. XXIX   b. XXXV   c VL   d. IL

Exercise 4b

1. a. 13 years   b. 9 years   c. 14 years
2.

Teacher	Age	Age in Roman numerals
Mr Austrie	44 years	VIL
Miss Francis	24 years	XXIV
Mr Jno Baptiste	43 years	VIIL
Ms Dailey	54 years	LIV
Mr Joseph	23 years	XXIII
Miss Laville	31 years	XXXI

 Now, try it out   (1D)

Students to work in teams of five and follow instructions to complete the exercise

Exercise 5a

1.

Numbers	Factors	Odd number	Even number	Prime number
23	1, 23	✔		✔
48	1, 2, 3, 4, 6, 8, 12,16, 24, 48		✔	
67	1, 67	✔		✔
70	1, 2, 5, 7, 10, 14, 35, 70		✔	

2. 2
3. 11, 13, 17, 19, 23, 29, 31, 37

Exercise 5b

1. a. 2, 4, 6, 8, 10, 12, 14, 16, 18, 20, 22, 24, 26, 28, 30, 32, 34, 36, 38, 40, 42, 44, 46, 48, 50
   b. 1, 3, 5, 7, 9, 11, 13, 15, 17, 19, 21, 23, 25, 27, 29, 31, 33, 35, 37, 39, 41, 43, 45, 47, 49
   c. 9, 18, 27, 36, 45
2. Any six of 53, 59, 61, 67, 71, 73, 79, 83, 89, 97
3. 56, 63, 70, 77, 84, 91, 98
4. 96, 108, 120, 132

Now, try it out   (1E)

Students to develop their own chart and work through the activity

Exercise 6a

1. a. 9   b. 15   c. 3
2. 1 and 2
3. a. 12, 24   b. 9, 18   c. 8, 16

Exercise 6b

1. a. 2 × 2 × 2 × 3   b. 2 × 2 × 3 × 3
   c. 3 × 3 × 5   d. 2 × 2 × 3 × 5

2.

Number	Factors
30	1, 2, 3, 5, 6, 10, 15, 30
44	1, 2, 4, 11, 22, 44
50	1, 2, 5, 10, 25, 50
75	1, 3, 5, 15, 25, 75

 Now, try it out   (1F)

14

Exercise 7a

1. a. 60   b. 150   c. 42
2. E.g. 6, 12, 18, 24, 30
3. E.g. 24, 48

Exercise 7b

1. a. 12   b. 30   c. 24   d. 18
2. a. 4, 8, 12, 16, 20   b. 6, 12, 18, 24, 30
   c. 9, 18, 27, 36, 45   d. 12, 24, 36, 48, 60
3.

Number	First five multiples
6	6, 12, 18, 24, 30
8	8, 16, 24, 32, 40
10	10, 20, 30, 40, 50
12	12, 24, 36, 48, 60

 Now, try it out   (1G)

E.g. 48, 96, 144, 192

Exercise 8a

1. a. 25   b. 144   c. 121   d. 100
2. a. 8   b. 4   c. 6
3. a. 5   b. 12   c. 8

Exercise 8b

1.

Number	Representation	Value
9^2	9 × 9	81
3^2	3 × 3	9
5^2	5 × 5	25
8^2	8 × 8	64
12^2	12 × 12	144
11^2	11 × 11	121

2. a. 49   b. 64   c. 36

 Now, try it out   (1H)

289

Exercise 9a

1. a. Kings Hill 2054 m < View Point 4534 m
      Belle Vue 4566 m > View Point 4534 m
      Belle Vue 4566 > Mount Hillaby 1112 m
2. a. −12 < − 2   b. − 115 < 0   c. (16 + 3 − 9) < (14 × 1)

Exercise 9b

1. a. A < B   b. C > B   c. B < D

 Now, try it out   (1I)

56 724

PRACTICE TEST 1 – NUMBER IDEAS AND RELATIONS – *Let's see how much you know*

Section A

1. d. (six thousands)
2. a. (25 438)
3. c. (March)
4. a. (February)
5. c. (sixty-four thousand, three hundred dollars)
6. d. (seventy-two thousand and thirty-four dollars)
7. a. (fifty-four thousand dollars)
8. b. (300 000 + 40 000 + 6000 + 700 + 20 + 9)
9. b. (23 480)
10. a. (3500)
11. a. 500 900
12. c. (LVI)
13. b. (46)
15. d. (8)

Section B

1. a. $45 680   b. $45 700
2. a. 1, 2, 4, 5, 8, 10, 20, 40   b. 24   c. 4
3. a. 20 000 + 3000 + 500 + 70 + 8
   b. 100 000 + 90 000 + 5000 + 300 + 60 + 4
4. a. 36   b. 8

5. a. 2007
   b. one hundred and thirteen thousand, six hundred and forty-two
   c. 112 300   d. 86 990   e. 2008 and 2009

CHAPTER 2 OPERATIONS ON NUMBERS

Exercise 1a

1. a. Month 2   b. 435 units   c. 199 units
2.

	Problem	Estimated answer	Calculated answer
a.	243 + 126 + 60	400	429
b.	1562 + 483 + 217	2300	2262
c.	245 + 314 + 723	1200	1282
d.	8794 − 2613	6200	6181
e.	2603 − 768	1800	1835

3. $110
4. 828 m
5. 4000

Exercise 1b

1. a. 5601   b. 1391   c. 604   d. 10 198
2. $600.00
3. a. 275 and 78   b. 78 and 7
4. $6423

 Now, try it out   (2A)

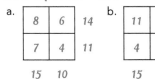

a.
8	6	14
7	4	11
15	10	

b.
11	8	19
4	12	16
15	20	

Exercise 2a

1. $385.00
2. 1632
3. 36
4. a. 4012   b. 42
5. 170 cm

Exercise 2b

1. 74
2. 812
3. 8 years
4. a. 36   b. 45   c. 78

 **Now, try it out   (2B)**

9 × 5 = 45
90 × 5 = 450
900 × 5 = 4500
9000 × 5 = 45 000
90 000 × 5 = 450 000

Exercise 3a

1. a. 279 000   b. 84.62   c. 49 200   d. 49 200
   e. 20 592   f. 64 135   g. 34 750   h. 4250

Exercise 3b

1. a. 320   b. 320   c. 4

Now, try it out   (2C)

1. a. 40   b. 100   c. 10

Exercise 4a

1. b. (17 + 32) + 96 = 145
   c. 17 + (32 + 96) = 145
   d. (17 + 96) + 32 = 145
2. Yes. Multiplication is commutative.
   a. no   b. yes   c. no

Exercise 4b

1. a. addition is commutative
   b. identity element leaves original number unchanged
   c. addition is associative
   d. identity element leaves original number unchanged
   e. number and its inverse sum gives identity
   f. multiplication is distributive over addition
   g. multiplication is commutative
2. a. (8 + 4) × 3 = (8 × 3) + (4 × 3)
   b. 121 × 1 = 121
   c. 302 × 12 = 12 × 302
   d. 5697 + 0 = 5697
   e. 16 + 0 = 16
   f. 36 + 73 = 73 + 36

Now, try it out   (2D)

A	B	A + B	A × B
65	66	131	4290
34	25	59	850
7	72	79	504

Exercise 5a

1. a. 11   b. 9   c. 42   d. 61   e. 20
2. a. 5500 − 1286 = 4214 and 5500 − 4214 = 1286
   b. 14 × 9 = 126 or 9 x 14 = 126
   c. 6336 ÷ 132 = 48 or 6336 ÷ 48 = 132
3. $2000

Exercise 5b

1. a. 6   b. 70   c. 14   d. 60   e. 2
2. 9
3. a. 6 × 68 = 408   b. 7 × 48 = 336   c. 8 × 82 = 656

 **Now, try it out   (2E)**

1. a. E.g (9 + 7) − (5 × 3) − 1 = 0
   b. E.g (3 + 1) × 5 × 7 × 9 = 1260

PRACTICE TEST 2 – OPERATIONS ON NUMBERS – *Let's see how much you know*

Section A

Select the letter of the correct answer to each question.

1. b. (take away 28)
2. a. (add 30)
3. d. (subtract 12)
4. a. (5 + 1 = 1 + 5)
5. a. (2880)
6. c. (767)
7. c. (product)
8. d. (0)
9. c. (1000)
10. c. (Division is the inverse of multiplication)
11. d. (9 and 7)
12. d. (22 000 × 1000)
13. b. (13 × 100) – 13
14. b. (30 × 12) + (30 × 3)
15. b. (60 students)

Section B

1. 108
2. 33
3. $9191.00
4. 3
5. 20

CHAPTER 3 WORKING WITH FRACTIONS

Exercise 1a

1. a. $\frac{5}{8}$ b. $\frac{9}{20}$ c. $\frac{4}{12}$ d. $\frac{6}{8}$
2. Correct fraction to be shaded:
   a. $\frac{6}{9}$ b. $\frac{2}{7}$ c. $\frac{7}{15}$ d. $\frac{4}{5}$

Exercise 1b

1. a. $\frac{7}{30}$ b. $\frac{5}{30}$ or $\frac{1}{6}$ c. $\frac{1}{30}$ d. $\frac{8}{30}$ or $\frac{4}{15}$
2. a. $\frac{5}{10}$ or $\frac{1}{2}$ b. $\frac{6}{8}$ or $\frac{3}{4}$ c. $\frac{1}{3}$ d. $\frac{5}{12}$

 **Now, try it out   (3A)**

Students to draw their own shapes and shade selected portions

Exercise 2a

1. a. Students to shade $\frac{7}{9}$, $\frac{11}{12}$, $\frac{33}{35}$, $\frac{96}{100}$
   b. Students to circle $\frac{12}{7}$, $\frac{18}{8}$, $\frac{35}{30}$, $\frac{56}{7}$, $\frac{25}{6}$, $\frac{95}{89}$
   c. Students to draw a box around $1\frac{1}{2}$, $3\frac{4}{5}$, $4\frac{3}{4}$, $6\frac{3}{5}$, $2\frac{2}{5}$
2. a. $1\frac{2}{7}$ b. $1\frac{14}{15}$ c. $8\frac{1}{12}$ d. $16\frac{6}{10}$ or $16\frac{3}{5}$
3. a. $\frac{7}{2}$ b. $\frac{46}{3}$ c. $\frac{111}{5}$ d. $\frac{75}{4}$

Exercise 2b

1.

Improper fraction	$\frac{8}{7}$	$\frac{15}{9}$	$\frac{22}{7}$	$\frac{33}{8}$	$\frac{19}{7}$	$\frac{35}{6}$
Mixed number equivalent	$1\frac{1}{7}$	$1\frac{6}{9}$	$3\frac{1}{7}$	$4\frac{1}{8}$	$2\frac{5}{7}$	$5\frac{5}{6}$

 **Now, try it out   (3B)**

Students to write their own fractions

Exercise 3a

1. For example
   a. $\frac{2}{4}$, $\frac{4}{8}$, $\frac{8}{16}$ b. $\frac{16}{24}$, $\frac{32}{48}$ c. $\frac{2}{6}$, $\frac{3}{9}$
   d. $\frac{28}{70}$, $\frac{56}{140}$ e. $\frac{60}{72}$, $\frac{120}{144}$
2. Fractions to be shaded:
   a. $\frac{6}{10}$ b. $\frac{4}{16}$ c. $\frac{18}{24}$ d. $\frac{4}{20}$

Exercise 3b

1. b. $\frac{3}{4}$ c. $\frac{20}{120}$ d. $\frac{6}{9}$

 **Now, try it out   (3C)**

Students to write their own fractions

Exercise 4a

1. a. $\frac{7}{8} > \frac{4}{8}$ b. $\frac{16}{24} < \frac{18}{24}$
   c. $\frac{20}{24} > \frac{12}{24}$ d. $\frac{30}{90} < \frac{36}{90}$
2. $\frac{19}{20}$, $\frac{9}{10}$, $\frac{3}{5}$, $\frac{5}{10}$, $\frac{2}{10}$

Exercise 4b

1.

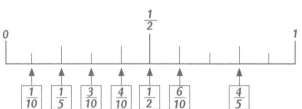

2. a. $\frac{7}{15} > \frac{2}{5}$ b. $\frac{3}{10} < \frac{4}{5}$
   c. $\frac{9}{30} < \frac{8}{15}$ d. $\frac{3}{4} < \frac{11}{12}$
3. The bird with $\frac{4}{5}$ carries the most.
4. a. $\frac{3}{8}$, $\frac{2}{5}$, $\frac{3}{4}$ b. $\frac{5}{8}$, $\frac{3}{4}$, $\frac{10}{12}$
   c. $\frac{11}{30}$, $\frac{4}{5}$, $\frac{5}{6}$ d. $\frac{1}{6}$, $\frac{2}{7}$, $\frac{2}{3}$

 **Now, try it out   (3D)**

$\frac{4}{24}$, $\frac{10}{24}$, $\frac{12}{24}$, $\frac{15}{24}$, $\frac{16}{24}$, $\frac{18}{24}$, $\frac{20}{24}$

Exercise 5a

1. a. $\frac{1}{2}$ b. $\frac{3}{8}$ c. $\frac{1}{2}$ d. $\frac{1}{3}$ e. $\frac{1}{4}$
2. $\frac{1}{5}$, $\frac{1}{4}$, $\frac{3}{10}$, $\frac{3}{8}$, $\frac{1}{2}$, $\frac{3}{4}$

Exercise 5b

1. 4 and 5
2. $\frac{14}{70}$ or $\frac{1}{5}$
3. $\frac{1}{9}$
4. $\frac{8}{9}$
5. The $4\frac{9}{16}$ kg bar

 Now, try it out   (3E)

Denominator 4
Dividing 5
Equivalent fractions 1
Fraction 2
Improper fraction 9
Multiple 10
A number 8
Numerator 7
Proper fraction 6
Thirds 3

Exercise 6a

1. a. Joey   b. $13\frac{9}{20}$ kg   c. $101\frac{1}{12}$ kg
   d. $145\frac{1}{10}$ kg   e. Sam
2. $\frac{1}{24}$
3. $\frac{3}{20}$

Exercise 6b

1. a. $\frac{3}{8}$ b. $\frac{1}{8}$ c. $\frac{3}{16}$
   d. Tessa   e. $\frac{16}{16}$ or the whole
2. a. $\frac{1}{6}$ b. $\frac{5}{6}$

 Now, try it out (3F)

Students to write their own fractions.

Exercise 7a

1. $\frac{3}{10}$
2. $37\frac{1}{2}$ hours
3. a. $\frac{7}{16}$ b. $\frac{3}{8}$ c. $6\frac{3}{5}$
   d. $\frac{24}{49}$ e. $29\frac{1}{6}$ f. $2\frac{13}{16}$

Exercise 7b

1. $3\frac{1}{8}$
2. $4\frac{1}{8}$ miles

3.

Fraction	Fraction multiplied by 2	Fraction multiplied by $6\frac{1}{2}$
$7\frac{1}{2}$	15	$48\frac{3}{4}$
$3\frac{4}{6}$	$7\frac{1}{3}$	$23\frac{5}{6}$
$1\frac{3}{4}$	$3\frac{1}{2}$	$11\frac{3}{8}$
$\frac{6}{8}$	$1\frac{1}{2}$	$4\frac{7}{8}$

Now, try it out   (3G)

Students to insert their own fractions

Exercise 8a

1. a. $\frac{2}{3}$ b. $4\frac{1}{8}$ c. $1\frac{1}{16}$ d. $1\frac{31}{32}$
2. 60
3. 15

Exercise 8b

1. a. 2 times faster
   b. 3 times faster
   c. The Mazda and the Nissan
2. Terry jumped $1\frac{29}{39}$ further than Bradly
3. a. Pam threw $1\frac{11}{39}$ further than Susan
   b. Pam threw 2 times further than Florence

Now, try it out   (3H)

Students to use their own fractions

PRACTICE TEST 3 – FRACTIONS

Let's see how much you know

Section A

1. b. $(\frac{1}{2})$
2. a. $(\frac{1}{3})$
3. c. $(\frac{1}{3})$
4. a. $(\frac{1}{4})$
5. d. $(\frac{9}{40})$
6. d. $(\frac{1}{4})$
7. b. $(\frac{15}{20})$
8. c. $(\frac{3}{15})$
9. c. $(\frac{75}{100})$
10. a. $(\frac{3}{4})$
11. b. $(\frac{1}{8})$
12. d. $(\frac{1}{16})$
13. a. (Randy received $\frac{1}{8}$ more than Mark)
14. c. (Tim)
15. a. ($3\frac{3}{8}$ boxes)

Section B

1. $\frac{3}{4}$
2. $\frac{3}{8}$
3. a. $5\frac{1}{4}$ b. $1\frac{1}{2}$
4. a. $\frac{3}{16}$ b. $\frac{1}{6}$
5. a. $\frac{1}{2}$ b. $3\frac{1}{3}$

ANSWERS

CHAPTER 4 DECIMALS

Exercise 1a

1. a. zero point zero two three
   b. six point seven eight
   c. five point zero two three
   d. twelve point nine eight
   e. fifteen point three zero four

Exercise 1b

a. 1.5   b. 8.68   c. 0.003   d. 10.034   e. 11.1

 Now, try it out   (4A)

Students to write 5 of their own decimals which are less than 50

Exercise 2a

1. a. 0.87   b. 1.43   c. 5.91   d. 7.51
2. a. 4.40, 4.53, 5.34, 5.4, 5.43
   b. 8.098, 8.10, 8.98, 9.105, 9.8
   c. 11.098, 11.34, 11.343, 11.45, 11.5
   d. 0.019, 0.023, 0.045, 0.17, 0.230

Exercise 2b

1. For example
   b. 7.35   c. 0.75   d. 5.0   e.12.05
2. a. >   b. >   c. >   d. >

Now, try it out   (4B)

Students to identify decimals of their own

Exercise 3a

1. a. $1\frac{1}{2}$ b. $\frac{8}{100}$ c. $15\frac{56}{1000}$ or $15\frac{7}{125}$
2. a. $\frac{2}{10}$
   b. four hundred and three point two zero six
   c. $403\frac{206}{1000}$ or $403\frac{103}{500}$

Exercise 3b

1.

Decimal	Fraction
12.67	$12\frac{67}{100}$
9.1	$9\frac{1}{10}$
5.004	$5\frac{4}{1000}$
6.302	$6\frac{302}{1000}$
3.5	$3\frac{1}{2}$

2.

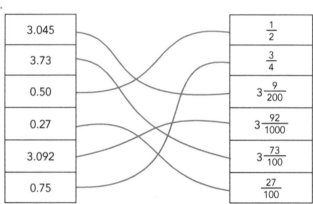

Now, try it out   (4C)

Students to identify decimals of their own

Exercise 4a

1. a. 0.16   b. 0.25   c. 1.41   d. 0.625
2.

Fraction	Decimal
$7\frac{3}{4}$	7.75
$15\frac{1}{4}$	15.25
$9\frac{1}{2}$	9.5
$12\frac{34}{100}$	12.34
$8\frac{1}{2}$	8.5

Exercise 4b

1. 5.25
2. 2.5
3. $\frac{1}{3}$ = 0.33 $\frac{1}{2}$ = 0.5 $\frac{1}{6}$ = 0.16

 Now, try it out   (4D)

Students to identify fractions of their own

Exercise 5a

1. a. 0.9   b. 7.7   c. 48.1
2. a. 46.6   b. 47
3.

Decimal	To the nearest tenth	To the nearest whole number
20.75	20.8	21
15.07	15.1	15
12.30	12.3	12
1.06	1.1	1
3.13	3.1	3

Exercise 5b

1. a. 66.8   b. 67
2. a. $46.70   b. $47.00

3.

Heat number	Time (in seconds)	Time to the nearest hundredth of a second	Time to the nearest second
1	54.567	54.57	55
2	52.340	52.34	52
3	49.085	49.09	49
4	48.993	48.99	49
5	47.047	47.05	47

 Now, try it out (4E)

Students to identify decimals of their own

Exercise 6a

1. a. 185.05 m   b. 200.105 m   c. 52.71 m   d. 9.313m
2. 21.225 m
3. 0.518
4. a. 70.9131   b. 41.065
5. a. 235.226 sq km   b. 242.74 sq km   c. 2.77 sq km

Exercise 6b

1. a. 25.663 m   b. 15.872 m   c. 60.076 m
   d. 20.352 m   e. 5.645 m

 Now, try it out (4F)

Students to identify decimals of their own

Exercise 7a

1. a. 10.05 m   b. 15.75 m
2. a. 120.92 miles   b. 211.61 miles   c. 362.76 miles
3. $21.25
4. 428.16 seconds

Exercise 7b

1. $34.83
2. a. 8.2 metres
   b. 205 metres
3. a. 28.0884   b. 0.02   c. 0.9   d. 36.48

 Now, try it out (4G)

Students to identify decimals of their own

PRACTICE TEST 4 – DECIMALS

➤ *Let's see how much you know*

Section A

1. b. (thirty-six point zero zero nine)
2. a. (1.025)
3. c. (Christelle)
4. b. (Sandra)
5. c. (68.094 s)
6. d. (68.997 s)
7. c. (1.046 s)
8. c. (11.410)
9. a. (34.603)
10. c. (5 hundredths)
11. a. (4 units)
12. d. (0.66)
13. a. (Belle Mont)
14. c. (0.306 miles)
15. d. (27.68 inches)

Section B

1. a. twenty-three point zero three
   b. eight hundred and ninety-one point seven six two
   c. 52.036
2. $15\frac{75}{125}$
3. a. 74.7   b. 75
4. a. 39.542   b. 33.43   c. 213.4
5. a. 74.42   b. 56.28   c. 31

CHAPTER 5 PERCENTAGES

Exercise 1a

1. 75%   2. 40%   3. 70%   4. 0.35   5. 70%

Exercise 1b

1.

Decimal	Fraction	Percentage
0.05	$\frac{1}{20}$	5%
0.25	$\frac{1}{4}$	25%
0.10	$\frac{1}{10}$	10%
0.33	$\frac{1}{3}$	33%
0.66	$\frac{2}{3}$	66%
0.75	$\frac{3}{4}$	75%
0.2	$\frac{1}{5}$	20%
0.125	$\frac{1}{8}$	12.5%
1.25	$1\frac{1}{4}$	125%

 Now, try it out (5A)

Students to identify fractions of their own and complete the exercise

Exercise 2a

1. a. 62.5   b. 180   c. 75   d. 16
2. 40%
3. 80%
4. 60%

Exercise 2b

1. 120
2. a. 5   b. 8.1
3. 40%

4.

Car colour	Number of cars	Fraction	Percentage
Green	5	$\frac{1}{10}$	10%
Red	10	$\frac{1}{5}$	20%
Yellow	15	$\frac{3}{10}$	30%
Black	12	$\frac{6}{25}$	24%
White	8	$\frac{4}{25}$	16%

 Now, try it out (5B)

Students to write their own numbers and calculate 10% of the numbers chosen

Exercise 3a

1. 20%
2. $66\frac{2}{3}$% or 66.6%
3. 23% (23.07% acceptable).
4.

Original quantity	New quantity	Difference	Percentage decrease
50 cars	30 cars	20 cars	40%
100 students	80 students	20 students	20%
50 birds	40 birds	10 birds	20%
20 books	15 books	5 books	25%
5 pens	3 pens	2 pens	40%

Exercise 3b

1. 40%
2. a. 85%   b. 90%   c. 5%
3.

Original quantity	New quantity	Difference	Percentage increase
200	250	50	25
125	150	25	20
70	90	20	28.6
15	25	10	66.6
4	6	2	50

 Now, try it out

15 more, 100%

PRACTICE TEST 5 – PERCENTAGES

Let's see how much you know

Section A

1. d. (42%)
2. c. (32%)
3. c. (40%)
4. c. (0.62)
5. a. (40)
6. d. (60%)
7. a. (25%)
8. d. (60%)
9. c. (40%)
10. b. (25%)
11. d. (50)
12. a. (2.5 items)
13. a. (20%)
14. c. (26%)
15. a. (30%)

Section B

1. 80%
2. 2500
3. 16%
4. 14%
5. 42%

CHAPTER 6 MONEY

Exercise 1a

1. $575
2. a. $394.04   b. $17 489.10   c. $4325.65   d. $6208.58
3. 4 hundreds and 2 fifty dollar bills
4. Either five $20.00 bills and four $50.00 bills or ten $20.00 bills and two $50 bills

Exercise 1b

1. $46.75
2. a. $315.20   b. $380.65
3. $13.50

 Now, try it out (6A)

Answers may vary based on pupils' responses

Exercise 2a

1. a. The shop is offering 25% discount on the original price
   b. $3000   c. 75%
2. $1400
3. $285.00
4. $14.25
5. Frying pans = $36.00   Glasses = $16.00
   Blender = $112.00   Microwave = $360.00
   Coffee maker = $48.00   Jug = $16.00
   Sandwich maker = $60.00

Exercise 2b

1. $333.33
2. $200.00
3. 15%
4. Savings on each item:
   Skirt = $2.00   Shoes = $11.00
   DVD = $9.00   Blanket = $52.50

Now, try it out   (6B)

Students to estimate the cost of 6 items then work out the cost with a 15% discount

Exercise 3a

1. 12% (nearest whole number)
2. 20%
3. 25% profit
4. 16% (nearest whole number)

Exercise 3b

1. a. 100   b. 1000   c. 880   d. 200   e. 50
2. a. $80.00 discount   b. $170.00
3. a. $4125.00   b. $375.00

 Now, try it out   (6C)

Best buy: microwave with price $550.00 less 20% discount

PRACTICE TEST 6 – MONEY

- *Let's see how much you know*

Section A

1. c. (two hundred and ninety-seven thousand, five hundred and six dollars and one cent)
2. a. (three hundred and forty-five thousand, eight hundred and ninety-seven dollars and nine cents)
3. a. ($650 496.00)
4. c. ($1 600 000.00)
5. d. ($5550.00)
6. c. ($310.90)
7. b. ($80.45)
8. a. ($108.00)
9. a. (20%)
10. c. ($420.00)
11. c. (25%)
12. a. ($80.00)
13. a. ($5400.00)
14. b. (50%)
15. b. (25%)

Section B

1. 50%
2. $25.00
3. $97.50
4. $7200.00
5. $360.00

CHAPTER 7 RATIO AND PROPORTION

Exercise 1a

1. a. 3 in every 4 squares are yellow
   b. There are 3 yellow squares to every 1 white square
   c. The proportion of yellow to white squares is 3 to 1

2. a. Students to shade 3 in every 5 loops red
   b. Students to shade 1 in every 7 loops black
   c. Students to shade 2 in every 3 loops green
3. a. 16   b. 24   c. 24   d. 28
4. a. 2   b. 3   c. 4   d. 10

Exercise 1b

1. a. 20   b. 25   c. 27
2. a. 25   b. 28   c. 15
3. a. 2 blue, 3 red, 4 green
   b. 4 blue, 6 red, 8 green
4. a. 1500 g flour   b. 600 g sugar   c. 1500 g flour

 Now, try it out (7A)

Students to identify their own number and utilize their own ratios

Exercise 2a

1. 1:10
2. 1 g sugar

Exercise 2b

1. a. $\frac{3}{4}$ b. $\frac{2}{3}$ c. $\frac{5}{8}$ d. $\frac{2}{3}$ e. $\frac{2}{3}$ f. $\frac{1}{5}$
2. a. 16 plates   b. 80 plates   c. $62.50

 Now, try it out   (7B)

40 sweets, 80 sweets

Exercise 3a

1. $60.00
2. a. $30.00   b. 180 cm

Exercise 3b

1. a. 6 pounds, 15 pounds
   b. Pamela = 10 lbs, Sophie = 20 lbs, Keisha = 30 lbs
   c. Brother 1 = 10 pounds tuna, 8 pounds red fish
      Brother 2 = 20 pounds tuna, 16 pounds red fish
2. 25 litres

 Now, try it out   (7C)

Ratio shared = 1:2:3

Exercise 4a

1. $24.00
2. 3 days
3. a. 36 days   b. 24 days

Exercise 4b

1. a. 2 days   b. 1 day
2. a. 5 days   b. 50 days
3. a. 120   b. 240

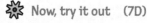 Now, try it out   (7D)

Students to follow instructions and complete the exercise

PRACTICE TEST 7 – RATIO AND PROPORTION – *Let's see how much you know*

Section A

1. d. (2)
2. a. (2:1)
3. c. (The ratio of blue to yellow shapes is 4:2)
4. a. (20 girls)
5. c. (24 boys)
6. a. (7 green sweets)
7. d. (1:2)
8. a. (1:4)
9. a. (30 towels)
10. c. ($50.00)
11. c. (6 days)
12. c. (10 minutes)
13. a. ($20.00)
14. b. ($40.00)
15. c. ($60.00)

Section B

1. 120 cm
2. 20 red sweets
3. 1 cm:100 miles
4. 1:4
5. 200 millilitres

CHAPTER 8 MEASUREMENT

Exercise 1a

2. 40 cm
3. 15 km

Exercise 1b

1. a. 4.05 km   b. 2.43 km   c. 1.62 km
2. a. centimetres or metres
   b. centimetres or metres
   c. kilometres
   d. centimetres
   e. millimetres
3. a. 14 750 metres   b. 14.75 kilometres

☀ Now try it out   (8A)

Students to measure objects indicated and answer the questions

Exercise 2a

1. 8 bags
2. 14 bags
3. 1.54 tonnes

Exercise 2b

1. a. 150 g   b. 8.5 kg
2. a. grams   b. tonnes   c. kilograms   d. grams   e. grams
3. 10 800 grams

☀ Now try it out   (8B)

Students to identify items which can be measured in kg and estimate their masses

Exercise 3a

1. a. 30°C = 86°F
   b. 60°F = 15.6°C
   c. 90°F = 32.2°C
2. a. Wednesday   b. Friday   c. 8°C   d. 12°C   e. 8°C

Exercise 3b

1. 10°C
2. 69°C
3. –8°C

☀ Now try it out   (8C)

Students to record temperatures and work out differences

Exercise 4a

1. 664 metres
2. 300 sq metres

Exercise 4b

1. 33.2 metres
2. a. 28 m   b. 16.5 m   c. 24 m   d. 28 m   e. 23 m

☀ Now try it out   (8D)

Students to identify rectangular objects in the classroom and estimate and measure dimensions

Exercise 5a

1. 900 cm^2
2. 8 cm
3. $165.00
4. a. 25 cm^2 b. 144 m^2

Exercise 5b

1.

Figure	Area of coloured triangle	Area of entire shape
a.	32 mm	64 mm
b.	18 mm	36 mm
c.	24 mm	48 mm

2. a. 20 cm^2 b. 13 cm^2
3. a. 24 m^2 b. 27 m
4. a. 7 m   b. 49 m^2 c. 24.5 m^2

☀ Now try it out   (8E)

1. a. 144 m^2 b. 25 m^2 c. 4 m   d. 12 cm^2
2. $275

Exercise 6a

1. Students' drawings
2. a. 31.4 cm   b. 78.5 cm^2
3. 188.4 cm

Exercise 6b

1.

Radius	Diameter	Circumference (2πr)	Area (πr^2)
2 cm	4 cm	12.56 cm	12.56 cm^2
5 cm	10 cm	31.4 cm	78.5 cm^2
10 cm	20 cm	62.8 cm	314 cm^2
12 cm	24 cm	75.36 cm	452.16 cm^2

2. a. 15.71 cm   b. 37.71 cm   c. 44 cm
3. a. 13.2 cm   b. 25.77 cm
   c. 13.86 cm^2 d. 52.83 cm^2

Now, try it out   (9C)

Students to follow the instructions and do the practical activity

Exercise 7a

1. 216 cm³
2. 30 days
3. 1500 ml or 1.5 litres

Exercise 7b

1. 3.375 litres
2. 1 litre

 Now try it out   (8G)

a. 13 500 ml
b. 14 $\frac{1}{4}$ litres or 14.25 litres
c. 11 litres

Exercise 8a

1. a. 18 000 seconds   b. 120 mins = 7200 seconds
   c. 72 hours = 4320 minutes   d. 444 minutes
2. $\frac{1}{12}$
3. 3600
4. 4:35 or 25 minutes to 5
   12:30 or half past 12
   9:50 or 10 minutes to 10
   2:40 or 20 minutes to 3

Exercise 8b

1. a. Monday   b. Friday   c. 8 hrs 40 mins
   d. 8 hrs 20 mins   e. 55 mins   f. 35 mins

 Now try it out   (8H)

60 minutes × 24 hours × 7 days = 10 080 minutes

Exercise 9a

1. a. 07:40   b. 17:35   c. 23:23   d. 00:00   e. 09:53
2.

Time on 12-hour clock	Time on 24-hour clock
9:59 am	09:59
5:45 pm	17:45
7:05 pm	19:05
7:12 am	07:12
3:05 pm	15:05

Exercise 9b

1. 19:10 h   3. 07:50 h
2. 23:45 h   4. 19:09 h

 Now try it out (8i)

Students to select and write their own times using 12- and 24-hour clocks

Exercise 10a

1. 9:05 am
2. 12 weeks 2 days
3. a. Jeffrey   b. 1 hr 30 mins
4. a. 1 hr 30 mins   b. English   c. 6 hrs 35 mins

Exercise 10b

1. 25 hrs late or 1 day and 1 hr
2. 3:42
3. 13 hrs
4. 578 mins
5. 10:00 am

 Now try it out (8)

second, month, decade, minute, century

Exercise 11a

1. 8 hours
2. a. 200 km   b. 5 hours   c. 50 km/h
3. 1 hour
4. 20 km

Exercise 11b

1.

Average speed	Distance	Time
60 mph	180 miles	3 hrs
24 mph	84 miles	3 $\frac{1}{2}$ hours
45 mph	90 miles	2 hrs
120 mph	120 miles	1 hr

2. 1100 miles
3. 2000 mph

 Now try it out

8 hours

PRACTICE TEST 8 - MEASUREMENT -

Let's see how much you know

Section A

1. c. (ribbon tied around a gift box)
2. c. (5 cm)
3. d.
4. a. (1 kg)
5. b. (40 m)
6. a. (624 quarts)
7. d. (324 cm²)
8. d. (88 metres)
9. a. (7 kg)
10. d. (1 tonne)
11. a. (30 metres)
12. a. (1 litre 750 ml)
13. d. (2 hours 35 minutes late)
14. a. (85 minutes)
15. a. (432 km)

Section B

1. 1080 cm³
2. Micah
3. Tom
4. 20 cuts
5. 26 °C

CHAPTER 9 GEOMETRY AND SPATIAL SENSE

Exercise 1a

1. Students' drawings

2. a.

Acute	Obtuse	Reflex	Right angle
JAB	ABC	CDE	TRS
HIJ	EFG	DEF	
BCD	RMN	FGH	
AQR	STR	GHI	
RST	IJA		

3. a. right   b. obtuse   c. acute

Exercise 1b

1. 8
2. Students' drawings
3. Students' drawings

 **Now, try it out   (9A)**

Students to draw their own angles

Exercise 2a

1.

Shape	Number of sides	Number of angles
triangle	3	3
square	4	4
rectangle	4	4
pentagon	5	5
hexagon	6	6

2. a. a square has four equal sides; a rectangle has two pairs of opposite sides equal
   b. rectangle and square
   c. rectangle, square and hexagon

3. any three from square, rectangle, kite, trapezium, rhombus, parallelogram

Exercise 2b

1. equilateral, isosceles, scalene
2.

Number of triangles	Number of rectangles	Number of squares
16	0	3

 **Now, try it out   (9B)**

Students to create their own patterns using the 2D shapes provided

Exercise 3a

1.

Name	Number of faces	Number of vertices	Number of edges
cuboid	6	8	12
cube	6	8	12
cylinder	3	0	2
cone	2	1	1
sphere	1	0	0
triangular pyramid	4	4	6

2. Students' own responses

Exercise 3b

1. a. cuboid   b. sphere   c. cylinder   d. cube
2. Students' own examples

 **Now, try it out   (9D)**

a. circle   b. triangle   c. rectangle

Exercise 4a

1. first pattern will make open cube
2. a. triangular pyramid   b. square pyramid   c. cylinder

Exercise 4b

1. Students' drawings
2. Students' drawings

Now, try it out   (9E)

Students to copy diagram and cut out the shape as indicated

Exercise 5a

1. B and K; G and E

Exercise 5b

1. Students' drawings
2. a cone

Now, try it out   (9F)

Students to draw and cut out their own shapes to complete the exercise

Exercise 7a

1. a. 4   b. 2   c. 1
2. a. no   b. yes   c. no

Exercise 7b

1 a. H, O, D, A   b.

Now, try it out   (9G)

1. Students' own drawings

2.

Number of sides	3	4	5	6	7	8	n
Number of lines of symmetry	3	4	5	6	7	8	n

PRACTICE TEST 9 – GEOMETRY AND SPATIAL SENSE
■ *Let's see how much you know*

Section A

1. c.
2. a.
3. c.
4. b.
5. a.
6. a.
7. d. (8)
8. b. (obtuse angle)
9. b.
10. d. (rhombus)
11. b. (60°)
12. d. (corresponding sides and angles equal)
13. b. (Sphere)
14. a.
15. a. (one)

Section B

1. Students' drawings
2. Students' drawings
3. 180°
4. 90°
5. Students' drawings

CHAPTER 10   DATA HANDLING AND PROBABILITY

Exercise 1a

1. a.

Month	Frequency	Total
Jan	IIII	4
Feb	⊪⊩	5
Mar	IIII	4
Apr	II	2
May	⊪⊩ I	6
June	⊪⊩	5
July	⊪⊩ III	8

b.

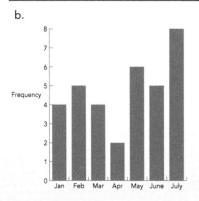

2. a. Students' responses
   b. Students' responses
   c. Students' own survey

Exercise 1b

1. a. A 9, E 8, I 11, O 6, U 1
   b.

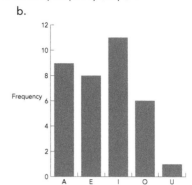

2.

Number	1	2	3	4	5	6
Frequency	2	2	1	4	3	4

3. a. USA   b. St Lucia   c. 30 students
   d. 20 students   e. 75 students

✸ Now, try it out   (10A)

Students to undertake a survey then construct a frequency table

Exercise 2a

1. a. $750.00   b. $1500   c. $300
2. a. Saturday 140 tiles, Sunday 80 tiles, Monday 60 tiles, Tuesday 160 tiles
   b. 440   b. 80
3. 31 cm
4. a. Antigua and Barbuda   b. Montserrat
   c. 13 000   d.

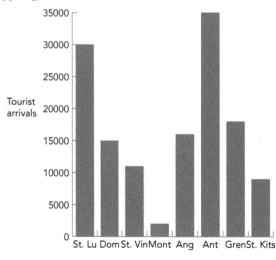

Exercise 2b

1. a. meat
   b. sweets
   c. fish
   d. 52 students
   e.

Food choices of students in a class	Number of students
Fish	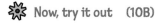
Sweets	☺☺☺☺☺☺
Cake	☺☺☺
Meat	☺☺☺☺☺☺☺☺☺☺
Fruits	☺☺☺☺☺
KEY ☺ represents 2 students	

2. a. (i)

Vehicle make	Frequency
Toyota	6
Mitsubishi	5
Honda	4
Nissan	7
Mazda	1
Suzuki	2

(ii)

Vehicle type	Frequency
Car	9
Van (4WD)	9
Bus	8

   b. Nissan   c. Honda   d. Toyota   e. Toyota
3. For example:
   • There was a drop in attendance from months April to May.
   • The greatest increase in attendance was seen from months May to June.

☀ **Now, try it out** (10B)

Students to construct their own pie chart using colour codes to represent the information

Exercise 3a

1. a. mean 36   median 30   mode 30   range 60
   b. mean 68   median 69   mode 75   range 21
   c. mean 2.25   median 2   mode 1   range 3
   d. mean 6   median 6   mode 3   range 7
2. a. mode 45 and 90, median 68, mean 67.4
   b. 55

Exercise 3b

1. a. Franklyn 9.5 feet, Brian 8 feet, Ricky 6 feet and 7 feet, Amos 7 feet
   b. Franklyn 8.6 feet, Brian 8 feet, Ricky 7 feet, Amos 7 feet
   c. Franklyn 8.59 feet, Brian 7.86 feet, Ricky 7.14 feet, Amos 7.64 feet
   d. Franklyn 2.5 feet, Brian 3.5 feet, Ricky 2.5 feet, Amos 2 feet
   e. Students' answers

☀ **Now, try it out** (10C)

Students to find out the ages of students in their class then complete the exercise

Exercise 4a

1. a. $\frac{24}{50}$ or $\frac{12}{25}$ b. $\frac{26}{50}$ or $\frac{13}{25}$
2. $\frac{1}{5}$
3. $\frac{1}{6}$

Exercise 4b

1. a. $\frac{2}{3}$ b. $\frac{1}{2}$
2. a. 40   b. 8   c. 11   d. Students' own questions

☀ **Now, try it out** (10D)

Students' own responses

PRACTICE TEST 10 – DATA HANDLING AND PROBABILITY

– Let's see how much you know

Section A

1. b. (basketball)
2. d. (boxing)
3. d. (85)
4. a. (50)
5. d. (15)
6. b. (70)
7. c. (335)
8. b. (40)
9. c. (52.5)
10. a. (59)
11. c. (10)
12. d. (14)
13. a. (22)
14. d. (228)
15. c. (45.6)

Section 2

1. $200.00
2. $450.00
3. $550.00
4. rent and bills
5. $1950.00